SMART GROWTH AND SUSTAINABLE TRANSPORT IN CITIES

This book delves into the urban planning theory of "smart growth" to encourage the creation of smart cities, where compact urban spaces are optimized to create transit-oriented, pedestrian- and bicycle-friendly areas, with a clear focus on developing a sustainable, humanistic transport system.

Over the last century, increased demographic changes and use of motor vehicles in the wake of "urbanization" led to the rapid expansion of cities, giving rise to economic, social and environmental problems. Sprawls and extension into natural areas caused a scattered urban context replete with empty spaces. This book provides an effective solution to this with an overview of the historical application of smart growth principles as a response to the issue of sprawling cityscapes, and sheds light on the theoretical information and methodologies used by cities to re-develop the urban landscape. It also encloses a checklist for practitioners and decision makers to inform the developmental process and integrate smart growth strategies into land use planning.

This book effectively engages with the global problem of urban sprawl in cities and hence will be an asset to both urban planning professionals, and graduate and postgraduate students of urban studies and the related disciplines.

Amir Shakibamanesh is an urban designer with more than 12 years of professional experience. He currently works as an associate professor of urban design at the University of Art, Tehran, Iran. He is the author or co-author of more than 25 scholarly articles and monographs, and the author of 8 books and book chapters. His research interests include urban modelling, urban simulation, virtual reality and urban scene analysis.

Mahshid Ghorbanian is an urban designer and planner with more than 10 years of professional experience. She currently works as an assistant professor of urban design and planning at the Iran University of Science and Technology, Tehran, Iran. She is the author or co-author of more than 14 scholarly articles and monographs, and the author of 8 books and book chapters. Her research interests include urban modelling, urban spatial structure and health, morphological and visual analysis of city textures, and urban transportation in the complex city network.

Seyed Navid Mashhadi Moghadam is a PhD candidate at the Faculty of Art and Architecture at Tarbiat Modares University in Iran. His research focuses on social aspects and dynamics in power distribution between citizens and governance.

SMART GROWTH AND SUSTAINABLE TRANSPORT IN CITIES

Amir Shakibamanesh, Mahshid Ghorbanian and Seyed Navid Mashhadi Moghadam

LONDON AND NEW YORK

First published 2020
by Routledge
2 Park Square, Milton Park, Abingdon, Oxon OX14 4RN

and by Routledge
52 Vanderbilt Avenue, New York, NY 10017

Routledge is an imprint of the Taylor & Francis Group, an informa business

British Library Cataloguing-in-Publication Data
A catalogue record for this book is available from the British Library

Library of Congress Cataloging-in-Publication Data
A catalog record for this book has been requested

ISBN: 978-0-367-26223-5 (hbk)
ISBN: 978-0-367-26224-2 (pbk)
ISBN: 978-0-429-29209-5 (ebk)

Typeset in Bembo
by Apex CoVantage, LLC

CONTENTS

FIGURES

TABLES

PREFACE

During the last century, the increasing demographic changes under the phenomenon of "urbanization" and the expansion of various types of motor vehicles have become the main problems of urbanization. Not only has the rapid expansion of cities led to economic and social problems but this human factor also has an environmental impact, which is among the most challenging issues in cities and urban areas. Urban sprawl and expansion into agricultural and natural areas have caused the formation of a separate and scattered urban context outside the cities that are full of empty spaces. Sometimes the empty space gets filled over time, resulting in a dense volume of uncontrolled urban context, which shows the ineffectiveness of development. Therefore, municipal decision makers have to consider the consequences of this uncontrolled growth. The lack of a comprehensive and efficient perspective in the field of pedestrian-oriented transport in many cities in the world has led to the accumulation of problems resulting from car-oriented transport.

Thus, innovative theoretical approaches have shifted from car-oriented transport to the development of sustainable pedestrian-oriented transport networks and development in their surrounding environments. While trying to resolve the issue of mobility in cities, they use novel tools and technologies to improve quality in the design of cities and return to the authentic and human traditions regarding movement in spaces. Meanwhile, using the smart growth approach in indigenous urban areas is inevitable, and perhaps developing operational strategies based on this approach can help us take a step towards solving the problems of cities and prevent their transformation into unsustainable residential zones with a host of chronic problems. Such a procedure could pave the way for scientific decision making for municipal authorities and help them predict the consequences of their decisions.

To build a society with a unique sense of place and an emphasis on minimal use of cars is considered the most important goal of smart growth that helps achieve the creation of a healthy society in which citizens have the basic amenities of living and are able to understand their residential environment and its readability. The main goal of this book is the realization of the principles of smart growth in order to achieve sustainable and humanistic transport. Other objectives of this book, which are based on the smart urban growth approach, are as follows.

- Developing indicators of smart growth and investigating the influence of implementing the requirements of smart growth in the formation of sustainable urban areas based on international experiences;
- Pinpointing applicable purposes, strategies and policies towards the realization of smart growth principles, with a special emphasis on facilitating pedestrian-oriented and humanistic transport.

This book consists of five chapters.

In Chapter 1, titled "Smart growth", key theoretical issues related to smart growth (such as definitions, history and stages of development, features and benefits of smart growth, etc.) are investigated and the practical implications of this approach (e.g. tools, strategies and techniques for smart growth) are discussed.

In Chapter 2, "Smart growth vs. urban sprawl", these two concepts are reviewed and their differences are compared and contrasted in various cases.

Chapter 3, "A review of global experiences in evaluating urban development plans and policies based on smart growth", does just this and investigates the realization of smart growth in some parts of the world.

Chapter 4, "A review of critiques of smart growth" discusses and evaluates some of the most important criticisms in this regard.

Finally, Chapter 5, "A comprehensive checklist of generalizable and achievable goals, strategies and policies for smart growth (with an emphasis on pedestrian-oriented transportation)", provides an exhaustive checklist of goals, strategies and generalizable executive policies that can be achieved based on smart growth, with an emphasis on pedestrian-oriented transport.

This book is adapted from some parts of research by the authors entitled "Creating Sustainable Urban Spaces Based on the Smart Growth Approach: Towards enhancing Urban Environment Design by Human-Based Sustainable Transportation (Case Study: District 12 of Tehran)" funded by Art University, Tehran, Iran, in 2017–2019.

INTRODUCTION

In both Europe and the United States, the surge of industry during the mid-to-late nineteenth century was accompanied by rapid population growth, unfettered business enterprise, great speculative profits and public failures in managing the unwanted physical consequences of development. Giant sprawling cities developed during this era, exhibiting the luxuries of wealth and the meanness of poverty in sharp juxtaposition. Eventually, the corruption and exploitation of the era gave rise to the Progressive movement, of which city planning formed a part. The slums, congestion, disorder, ugliness and threat of disease provoked a reaction in which an improvement in sanitation was the first demand. A significant betterment of public health resulted from engineering improvements in water supply and sewerage, which were essential to the further growth of urban populations.

In the twentieth-century city planners solved many urban issues, such as need for sanitation and the lack of urban green spaces; however, the expansion of cities under the guise of rapid economic developments caused new major problems. The uncontrolled and increasing expansion of cities has led to the depletion of natural resources and an increase in the harmful effects of urbanization. The irrational distribution of populations on the natural environment has degenerated nature and increased the economic costs of the inter- and intra-city infrastructure, jeopardizing the healthy communities required to place humans in a healthy environment.

Today, given the increase in the production of cars and the willingness to use personal cars, as well as the misplaced emphasis on the part of some decision makers in the field of urban management on fulfilling all needs of car transportation (such as providing maximum space for moving and parked vehicles and efforts to facilitate and expedite their movement in all urban spaces), a shift of emphasis towards sustainable urban transport systems is seriously felt in many countries.

In fact, in the current situation, a car-oriented lifestyle has dominated the planning and design approaches in cities and towns across the country, to the extent that meeting the needs of pedestrians in cities has degraded to a subsidiary issue, which often goes unnoticed. It seems that in today's cities, a car-oriented approach in urban areas has gone so far that even small and micro-scale urban trips, which often were done on foot less than two decades ago, have completely changed and in most cases been eliminated from the urban lifestyle. This has also led to a reduction in the health of the urban environment in two ways. First, the amount of physical activity of urban people has greatly reduced. Second, environmental pollution has been on the rise due to irrational use of cars.

The scholars in urban sciences seek to alleviate the negative effects of the growing population and scattered expansion of cities. To this end, they have attempted to provide efficient solutions to improve human transportation to reduce and eliminate the concerns of urban sprawl and horizontal expansion of cities. Indeed, lack of a serious approach to the creation of sustainable urban space and efforts to design an urban environment that gives priority to sustainable and humanistic transport prompted us to focus this book on taking advantage of efficient and applicable smart growth guidelines.

Smart growth, as one of the most practical views expressed in the final decade of the twentieth century, is rooted in the concepts of sustainable development. Stephen Plowden and Andres Duany, the pioneers of the theory, deem it an approach aimed at dealing with urban and suburban sprawl in order to achieve a unique sense of community and place. In fact, the idea of smart growth relies on the concept of sustainable development with an emphasis on the development of urban centres based on public transport and compact and mixed land use and neighbourhoods with high walkability potential to provide a range of residential choices to designers and urban planners.

The smart growth movement represents an important contribution from North American planning theory on the issue of curbing urban sprawl. Smart growth has been defined in many different ways (it will be discussed in the first chapter); nevertheless, a general consensus exists in considering it as part of the broader sustainable planning movement. Smart growth needs to be pursued in a complex framework of multi-faceted actions, by embracing a wide array of sectors, both in planning and policymaking fields, and by acting with a multi-level approach. In fact, smart growth is a combination of its prior theories of new urbanism, such as the garden city, compactness and sustainability.

The use of smart growth became popular in the mid-1990s, thanks to the effort of the Maryland governor, Parris Glendening, whose primary agenda was to turn the state's development pattern into a more sustainable one. In 1997, he proposed a specific piece of legislation aimed at conserving open spaces and discouraging the scattered and leapfrog development which was occurring. Furthermore, he led the creation of an initiative that focused on using the entire $23 billion state budget as an incentive for smart growth. Smart growth has now extended to many other institutional initiatives by state, regional and local authorities.

Smart growth allows urban communities to manage their growth in a way that supports economic development and jobs; creates strong neighbourhoods and a diverse range of residential, commercial and transport options; and creates healthy communities with favourable places for citizens. Smart growth provides solutions to solve the problems of communities that have experienced sprawl in the past 50 years. Smart growth tends to increase resource efficiency in several ways. It reduces per capita land consumption and impervious surface area. It improves accessibility and transport diversity, which reduces per capita motor vehicle travel and associated costs. Smart growth results in cleaner air, lower taxes, less gridlock, more green space, increased accessibility, more housing options and preserved farmlands.

The advocates of this theory stress development based on extensive public transport with limited environmental impact. They seek to create a variety of transport options, including pedestrian-oriented and cycling capabilities in neighbourhoods. Therefore, in large municipalities, business groups based in the central part of cities and private investors often prefer smart growth as a means to revitalize neighbourhoods and urban centres without negative effects on existing valuable environmental or social conditions.

In fact, smart growth is a common term for the integration of transport and land use in a way that supports compact development and mixed urban uses, in contrast to car-oriented and scattered development on the periphery of cities. Thus, policies and strategies for smart growth allow optimum use of land and regulate available land use patterns to promote the quality of sustainable access in urban spaces, provide more open and green spaces and thus reduce the environmental consequences of urbanization.

In order to create a society with a unique sense of place and with special emphasis on sustainable and humanistic transport, smart growth aims at creating a healthy, high-quality, responsive and perceptible environment. In fact, smart growth particularly focuses on improving the accessibility of urban spaces through improving walkability, as is evident in its principles. Therefore, this approach seeks to place the distribution of urban activities and functions that people often need in their vicinity. That is why the main unit of smart growth design is the neighbourhood unit, or local community. In the same line, "creating walkable neighbourhoods" is the fourth principle of smart growth. This book tries to incorporate smart growth principles and standards and remove its operational flaws and weaknesses to pave the way for the emergence of urban development and design principles in a way that can be applied on the ground.

1

SMART GROWTH

From theoretical approaches to practical concepts

An introduction to a necessity: what did urban sprawl do to the cities?

Urban sprawl is based on a low-density lifestyle – or the American Dream – and seeks easy access to open spaces, freedom of movement and escape from problems such as poverty and the bustle of downtown areas as the most important goals. In other words, urban sprawl is an auto-dependent land development often leapfrogging away from the current denser development nodes to undeveloped land and separates where people live from where they work which therefore requires cars (Gillham, 2002). Forty years ago, extensive studies were carried out in relation to the issues, problems and extent of urban sprawl – both in terms of quality and quantity (Mirowsky & Ross, 2015; Nazarnia, Schwick, & Jaegera, 2016; Squires, 2002; Vos, Acker, & Witlox, 2016). Snyder and Bird (1998) provide a good description of the effects of urban sprawl on the American way of life. They define urban sprawl as a suburban development based on low density in barren lands (Snyder & Bird, 1998).

In 1974, the Real Estate Research Corporation submitted a report to the U.S. Council on Environmental Quality, the Department of Housing and Urban Development and the Environmental Protection Agency and suggested the need for intensive studies on the negative impact of urban sprawl (The Real Estate Research Corporation, 1974). Some other studies introduced urban sprawl as the result of a combination of several factors, including the growth of urbanization; the loss of open, cheap lands from cities; the development of transportation systems; increased capital available to people to buy properties; increased land speculation and profiteering; mass production of housing; and access to single-family dwellings becoming an ideal for different social groups. This form of urban development, which is also called "external" or "peripheral" development, lacks spatial coherence and depicts some kind of urban illness (Burchell, Shad, Listokin, Phillips, & Downs, 1998; Carruthers & Ulfarsson, 2002; Litman, 2015b).

This type of unbridled urban development that has occurred principally on lands not prepared for such purposes has various consequences, including increased unused lands, increased share of open spaces, lower population density, social isolation, distinctive land use, lack of environmental variety, reduced attractiveness of landscapes, marginalization, shortage of public service and so on (Cho, 2008; Sroka, 2016). In addition, it accelerates the erosion of downtown areas and prevents allocation of new amenities to them. Thus, in the view of many experts, excessive urban development in the periphery of cities in the United States has led to the worst kind of urban structure (Gordon & Wong, 1999).

Today, with the proliferation of the horizontal expansion of cities, urban sprawl is investigated and evaluated in different countries using various measures. And because each of these countries has different features and contexts, urban sprawl will have different results (Galster, Hanson, Ratcliffe, & Woldman, 2001; Gillham, 2002). In the United States, as the origin of the concept, urban sprawl has become a subject of interest in urban planning in recent decades (Anacker, 2015; Ewing, Hamidi, & Grace, 2016a; Ewing, Hamidi, Grace, & Weid, 2016b). In the United States, the rate of transformation of lands to urban areas has been higher than the rate of increase in the urban population; therefore, the increase in the quantity of land use has not been solely a result of population growth (Ewing et al., 2016b).

The urban design presented in the nineteenth century in the form of ideal, theory and law, took greater form in the twentieth century. At this time, besides an emphasis on engineering, urban design addressed sanitation improvement and economic improvement and was seen as a sort of artwork on a city scale. However, following World War II and in the 1950s, governments became architects trying to rebuild the world. In the mid-twentieth century, comprehensive planning for slum clearance, urban renewal, public housing and infrastructure projects became the main urban planning practices (Chaolin, 1994).

The Green Revolution in late 1970 provided innovative strategies such as land preparation, expandable residential nuclei, suburban promotion and small investments for the improvement of buildings (Cleaver, 1982). With the rise of non-governmental and community-based organizations, decentralization and empowerment of local governments increased. On the other hand, the collapse of the Soviet Union showed the challenge of centralized reorganization and market-based models of urban development and housing and clarified the means of privatization, land transfer, etc. (Andrusz, Harloe, & Szelényi, 2011).

In the 1990s, with the advent of the New Urbanism movement, the compact form of traditional cities was once again deemed to be the core idea of urban development and praised by urban planners. New Urbanism widely criticized the twentieth-century suburban development strategies and supported the idea of redeveloping downtowns and older areas of cities, especially places damaged due to the concentration of poverty. Also, this movement stressed the expansion of mixed use that could help create high-quality urban places (Bohl, 2000).

New Urbanists deem unlimited growth, traffic congestion, racial contradictions and conflicts, recession and a worn-out infrastructure as the main problems of cities in the twentieth century and believe that in a time demanding a better quality of life, modernism has taken the lead in creating ugly uniform cities. Krieger (1991) states that suburbs have destroyed the cities and their open spaces. Their non-normative development of car-oriented spaces has caused serious damage to the cities and destroyed the aesthetics of space and sociability in urban neighbourhoods of small towns or original parts of cities of the nineteenth century (Krieger, 1991). New Urbanists depict as their ideal a city without suburbs, the inner parts of which are multi-functional and inter-connected and include places to live, work, shop and have fun.

New Urbanism opposes the rapid growth of an irregular physical-spatial structure of cities and rejects excessive attention to car-oriented forms of development that expands streets and parking lots, that in turn, challenge the commercial sector of cities. It stresses pedestrian-oriented spaces. Supporters of the movement try to return people and businesses back to the central parts of cities and put an end to what Robert Cervero (1996) calls locked suburbs.

Various methods have been used in urban planning to solve the widespread social, economic and ecological problems of urban sprawl. Infill development, compact cities and smart growth are the most important examples of such methods. Taking an approach that can control the city's irregular growth and present a model based on sustainability for addressing urban issues is of high importance. Figure 1.1 presents the consequences and adverse effects of urban sprawl in the form of a conceptual diagram.

The next section of this book briefly discusses two responses to urban sprawl: the compact city and transit-oriented development, and after that discusses whether smart growth can be an appropriate and comprehensive response to urban development. To answer this question, first smart growth is defined. After explaining smart growth and its boundaries, the relationship between smart growth and sustainable development and the direction of such a relationship is evaluated. In addition, in order to identify the desired state of cities from a smart growth perspective, the goals and objectives of this approach will be examined. After explaining the position of smart growth, the methods of achieving its intended purposes should be determined. To this end, the principles of achieving smart growth will be studied and analysed. At the end of this chapter, the effects of smart growth and the related benefits and costs will be discussed briefly.

Compact city

The rapid and horizontal sprawl and growth of modern cities has caused many environmental, economic and social problems in many countries. The American and Australian theorists defend this theory under the name of development decentralization, and its practical manifestations can be found in countries such as the United States, Australia and Canada, and includes sprawling suburbs (Bourne, 1992;

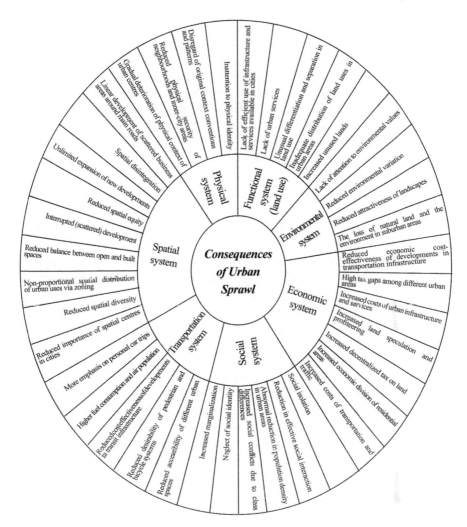

FIGURE 1.1 The consequences and effects of urban sprawl

Source: Authors

McManus, 2005). Although this urban development method represents cities with the most consumption of fossil fuels, the most harmful environmental impacts of pollutants and higher greenhouse gas emissions resulting from these fuels, due to the vast territorial area of these countries, the effects of pollutants and environmental degradation in these cities, are less likely to be seen as a concentrated issue (Campbell-Lendrum & Corvalán, 2007; Kennedy et al., 2009). In general, the most important features of the sprawling form of urban development are lower density, high dependence on vehicles, land use segregation, lack of biodiversity, reduced attractiveness of landscapes, excessive urban sprawl towards the outskirts and decentralized land ownership that, as a result of an increased share of open spaces and urban discontinuity, leads to reduced population density and social segregation (Ziegler, 2003).

The compact city approach was raised in opposition to the urban sprawl that emerged in the urban development process (Dantzig & Saaty, 1973). A brief study of the history of urban planning and its evolution and development clearly shows that scholars have presented different theories with different orientations in the field of urban sprawl (Burton, Jenks, & Williams, 2003). Peter Hall believes that urban planning in the twentieth century was raised in response to the problems and anomalies that emerged in the nineteenth century (Hall & Pfeiffer, 2013). Compact cities are part of the sustainable future of our cities. It is evident that there is a strong relationship between urban morphology and sustainable development. In general, it is suggested that in order to create a sustainable city, the forms and scales suitable for walking, cycling and public transportation should be connected to each other (Elkin, McLaren, & Hillman, 1991).

The compact city is a method to reduce travel distances. Thus, emissions of greenhouse gases and other pollutants are reduced, and ultimately, global warming can be avoided. By reducing the consumption of fossil fuels, residents of urban areas can pay less for transportation (Breheny, 1995).

In developing countries, compact development, which is the same concept as in developed countries, can lead to reduced travel distances, which in turn will reduce air pollution. High density provides viability through the provision of daily services; public transportation; reduced waste generation; and access to health, medical care and educational centres. Therefore, one of the methods to achieve such goals and to overcome existing problems is to change the shape of cities – how they are developed and how they are governed – that the compact form of cities provides in such a setting. In urban studies, the main concepts of this idea can be found in new approaches such as New Urbanism.

Public transit-oriented development

The public transport-oriented development model was proposed by the American architect Peter Calthrope based on the concept of the pedestrian packet, the philosophy of which is to facilitate spatial communication, and as a result, achieve high social solidarity (Calthorpe, 1989, 1994). By integrating transportation planning and land use, public transport-oriented development aims to prevent the urban sprawl and development toward the city outskirts to assume a mixed, compact and relatively high-density urban form (Zhao, 2010). This development is associated with residential and commercial centres and is designed based on maximum access to public or non-motorized transportation. Accordingly, a bus station or a rail transport is located at its centres and is surrounded by relatively high-density developments; up to a distance of half a mile (900 metres) outward, its density is reduced (Calthorpe, 1993). Also, in a comprehensive definition, Calthrope states that a public transport-oriented development is composed of a walkable neighbourhood located at a distance of 2,000 feet (600 metres) from a public transport station with commercial centres that have mixed land uses, such as residential, retail and office land uses and public spaces (Calthorpe, 1993).

By joining the founders and leaders of the New Urbanism movement, Peter Calthrope added the public transportation network to the traditional neighbourhood and, by creating public transportation–coordinated developments, expanding pedestrian routes and increasing the density adjacent to public transportation stations, he took a major step toward reducing the challenges in the suburban development model (Cervero, 2004; Renne, 2005). In fact, the public transit-oriented development, which is sometimes also called the transit village, represents a clear understanding of the close relationship between transportation and land use. This development offers an alternative to the car-dependent suburban living. One of its main features is settlements with higher density and mixed land uses. In most cases, denser centres are located within existing suburban areas to facilitate the use of public transport and make walking both comfortable and attractive (Cervero, 2004). As Cervero (2004) explained, in general, transit-oriented development places a higher density centre with mixed land use within a designed settlement, usually less than half a mile away from the transport station (which is suitable as the walking distance). The resulting developments are similar to those of the old railroad-oriented suburbs such as Forest Hill Gardens in New York or Evanston in the Chicago suburbs (Keating, 2008).

As defined, like the suburbs dependent on their roads and the urban rail transit, the transit-oriented developments are connected to a main line of the regional transportation network either directly or via a bus line (Bertolini & Spit, 2005; Schlossberg & Brown, 2004). However, its distance from the line of the public transport network should not require more than ten minutes of riding. Usually, these routes are connected to and coordinated with a larger regional strategy, which is the same as the public transit-oriented centre (Bertolini & Spit, 2005).

In fact, the combined uses, including residential, retail, employment, administrative, research, etc. form the basis of public transit-oriented development projects (Bertolini, 1996). A favourable public transport-oriented development goes beyond density and proximity. In fact, incentives and land use regulations ensure that public transit-oriented neighbourhoods are sustainable, because they are well designed, have good walkability and energy efficiency and have created the sense of place as an encouraging stimulus for social interactions (Suzuki, Cervero, & Iuchi, 2013). Therefore, what is referred to in all definitions concerning public transit-oriented development is:

- Planning for the area adjacent to the station and changing its role from being just a node to being a location;
- Creation of compact and densely populated neighbourhoods and blocks along with optimal pedestrian routes and land use combinations;
- Optimal increase of construction density and height (medium to high);
- Parking management;
- Increased quality and accessibility of public transportation;

- Promotion of non-motorized means of transportation such as cycling and walking;
- Good design of the routes and sites;
- Development of a network of public green and open spaces;
- Providing varied options for housing in terms of form, composition and price (Cervero, Ferrell, & Murphy, 2002; Dittmar & Ohland, 2012).

Some of the public transport-oriented development objectives can be introduced as:

- Meet the demand for travelling by public transport during peak and non-peak hours;
- Coordination between the transportation and the future land use models;
- Travel demand management (TDM);
- Concentrating the origin and destination of the trip in a single place (Dittmar & Ohland, 2012).

Major public transit-oriented development policies

The major public transport-oriented development policies include "organizing urban centres" and "controlling urban development in the suburbs." In general, regarding the general principles of this development, one can argue that this model should encourage walking and public transportation and reduce the use of cars. The common means of facilitating this development is the creation of places with design features such as combining walking paths with landscape design, locating parking behind the building and creating commercial streets that make walking and public transport more enjoyable.

The benefits of implementing public transit-oriented development

As one of the most prominent forms of smart development, public transport-oriented development is proposed as an antidote against the traffic congestion, isolationism and dispersion found in suburban societies; the lack of affordable housing; and reduced investments in urban areas. Public transport-oriented development strategies can provide a sustainable urban form for urban development. Also, under appropriate conditions, it can be considered an advantage for the local communities, especially when accompanied by active public participation (Cervero, 2004). Evidence suggests that in more compact development, people are 30 to 40 percent less likely to use their personal cars, which will lead to financial and health benefits (Cervero et al., 2002). According to the table, the public transport-oriented development benefits are grouped into several categories. Some of them are related to the general public, and some others are privately owned by specific individuals, property owners and businesses. Also, some of the benefits, such as the

TABLE 1.1 The benefits of implementing public transport-oriented development policies

Benefits	Primary beneficiaries of the public transport-oriented development benefits	
	Private sector	Public sector
Primary benefits	• Increased price of land and rent and increased efficiency of personal properties • Increasing the opportunities of constructing affordable housing	• Increased number of public transport passengers and reduced cost of travel • Providing development opportunities adjacent to public transportation • Revival of urban neighbourhoods and economic development
Secondary benefits	• The boom in retail • Increased access to employment places and work centres • Reduced cost of parking • Increased physical activity	• Reduced costs for traffic congestion and travel by personal vehicle, such as pollution and fuel consumption • Increased efficiency of retail and property taxes • Reduced urban sprawl and maintaining open spaces • Reduced road expenses and building the required infrastructures • Reduce crimes • Strengthening social capital and increasing public participation

Source: Cervero, 2004

opportunities for developing affordable housing, belong to both public and private sectors. Table 1.1 groups the benefits into two categories of primary and secondary benefits. The primary benefits are those that show a direct relationship between the public transport-oriented development and its effects, and the secondary benefits are defined based on the primary benefits and can also be achieved simultaneously (Cervero, 2004).

Disadvantages of public transit-oriented development implementation

Apart from the numerous advantages of this kind of development, many experts also refer to some of the disadvantages of this model. They claim that transit-oriented urban development is the product of the boutique-oriented urban development or nice-looking urban development mind-sets, which focus on physical design without paying due attention to social behaviours (Belzer & Autler, 2002; Currie, 2006; Litman, 2004). A lack of supportive legal frameworks, high initial investment costs and financial problems stemming from a small number of investors and developers

TABLE 1.2 The report of the California Department of Transportation

Factor	Description
Weak design of the existing public transportation system	Poor access for pedestrians and neglect of local communities in the area around the station, and their separation from the station by large parking lots and poor connection with activity centres are some of the most important obstacles to implementation.
Local communities' concerns	Concerns of local communities about identity change and neighbourhood structures and increased density and traffic, despite the local governments' support for public transport-oriented development, are the major obstacles to implementing this approach.
Non-compliance of zoning laws with the public transportation system	Increased local development around stations tends to be of low density and expands vehicle-oriented land use.
High cost and risk of compact development	Combined development and the increased density, coupled with smaller parking areas, can be a risk to investors and developers compared to the so-called vehicle-oriented approach to urban development; it is more complicated and costlier, with more rules.
Little investments made by the private sector	The fundraising by private investment is one of the main obstacles. Investors are less likely to pay attention to mixed land use licenses with smaller parking areas. Also, the private-sector investment for TOD is very small.

Source: California Department of Transportation, 2005

due to the uncertainty of its results are among the disadvantages of this model (Papa & Bertolini, 2015). There are also obstacles to the implementation of public transit-oriented development in urban areas that slow down the development process and can challenge it. In its report, the California Department of Transportation mentions the five factors listed in Table 1.2 as obstacles to the implementation of this type of development.

Obstacles to the implementation of public transport-oriented development

The Transit Cooperative Research Program, Report 102 (TCRP), Cervero (2004) states that public transport-oriented development implementation can be classified into three main categories, including financial barriers (those that reduce the ability to implement such projects), organizational and institutional barriers (such as structural barriers in the body of transport organizations and other governmental institutions responsible for the implementation of these projects) and strategy and policy barriers (such as land use and zoning policies) (Cervero, 2004, p. 99). They also believe that some of the barriers are specific to the strategy of public

TABLE 1.3 Obstacles facing the implementation of public transport-oriented development projects

The main obstacles	Description
Traffic congestion dilemma	The concentration of development in a single place always increases the traffic congestion and reduces the service level.
Logistical dilemma (due to the contradiction between node and place)	The moving bustle in stations, interference of different movements and indirect paths of motion with improper use of space are obstacles to creating a human-scale, comfortable place as part of a community. In fact, a contradiction occurs between the role of station location and the intersection function for access to the rail and bus transports.
Parking dilemma	In many developments, despite the availability of free parking spaces, building regulations emphasize construction in parking spaces per residential unit that increases the cost of the residential unit by $20,000. On the other hand, in some public transport-oriented development projects, the excessive restrictions on parking construction will lower the cost to below that of the market demand, and therefore, fewer investors are willing to do this.
Achieving the correct land use combination formula	Each piece of land with a specific ownership has a specific investor, a lender, a contractor and investment parameters, and combining these projects with each other seems to be problematic. Due to costs and various challenges, investors are less likely to resort to mixed developments and prefer the current mix of land uses in their vertical form.

transport-oriented development and are different from the obstacles mentioned earlier. These barriers are presented in Table 1.3.

Types of public transit-oriented development

Many experts argue that public transport-oriented development can cover a variety of shapes and station areas used on a different scale that complements the function within the system. In 1993, Peter Calthrope differentiated between the public transit-oriented development at the level of the mostly residential neighbourhoods and this type of urban development, which emphasizes employment-creating land uses (Calthorpe, 1993).

In 2012, Dittmar and Ohland redefined these distinctions and offered a variety of public transport-oriented development ideas that included the business centre of the city, urban neighbourhoods, suburban centres, suburban neighbourhoods and neighbourhoods in transitional areas and towns (Dittmar and Ohland, 2012). White and McDaniel (1999) also defined six forms of public transport-oriented development in different geographical contexts: 1- Single-function axes (such as office or retail); 2- Axes with mixed usage; 3- The neo-traditional development (focused on the features associated with the traditional villages and old cities); 4- Compact public transport-oriented development with mixed land uses around the transport station; 5- Rural context (emphasis on single-household dwellings around a central green space with a square); and 6- The peripheral region, which covers an area of about 150 acres (60 hectares) with a population of about 700 people that is subject to the rules of development plans but is faced with fewer functional limits (White & McDaniel, 1999). In a report presented by a public transport-oriented development centre, various types of public transit-oriented places are divided into eight general categories that cover the urban area to the developmental axes as shown in Table 1.4.

TABLE 1.4 Various types of public transit-oriented places

Scale	Description
Regional Centre	Regional centres are served by a rich mix of transit modes. For example, San Francisco is served by heavy rail, light rail, streetcar, cable car and high-quality bus. The entire half-mile radius around stations is dense, with the intensity increasing slightly in the quarter-mile radius.
Urban Centre	*Urban centres contain a mix of uses at slightly lower intensities than regional centres. They are commuter hubs to the larger region and are served by multiple transit options. Densities and intensities are usually greater in the quarter-mile radius of stations than in the half-mile radius.*
Suburban Centre	*Suburban centres act as both origin and destination for commuters, with a mix of transit options connecting to the regional network. Development is more recent than in urban centres, with more single-use areas and notably greater intensities in the quarter-mile radius of stations.*
Transit Town Centre	Transit town centres are local serving centres of economic and community activity served by a variety of transit modes, primarily providing commuter service to jobs in the region. Densities are usually noticeably greater in the quarter-mile radius of stations.
Urban Neighbourhood	*Urban neighbourhoods have moderate to high densities and transit is less of a focal point of activity than in centre place types. Intensities are usually spread evenly throughout the half-mile radius with an increase near the station.*
Transit Neighbourhood	*Stations in transit neighbourhoods are less a focus of activity than in the previous place types and usually do not have enough density to support much local-serving retail. They are typically served by rail or multiple bus lines at one location.*
Special Use/ Employment District	*Special use/employment districts are often focused around a university or sports stadium, and stations are not the focus of economic activity. Densities are usually evenly distributed in the half-mile radius around stations*
Mixed-Use Corridor	*Mixed-use corridors offer good opportunities for infill and mixed use development. They are a focus for economic and community activity but have no distinct centre, though development is usually more intense within a quarter mile of transit stops.*

Source: Reconnecting America's Center for Transit-Oriented Development, 2008

The definition of smart growth

The American Planning Association (APA) defines smart growth as follows:

> *"Smart Growth could see as a plan/design frame work to achieve a better development pattern, a unique sense of community and place, preserve and enhance valuable natural and cultural resources in metropolitan areas, suburbs and neighboring towns.*
> *(American Planning Association, 2002)*

Figure 1.2 shows the components of smart growth and the relationship among them based on the views of the APA.

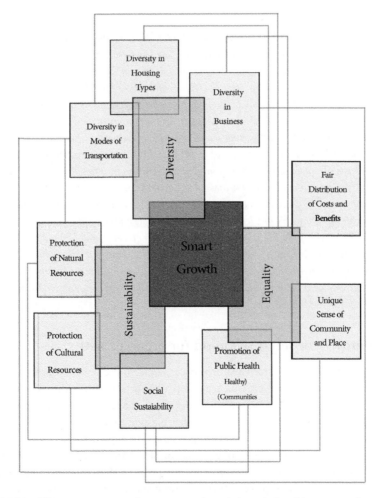

FIGURE 1.2   The components of smart growth and the relationship among them based on the views of the American Planning Association

Source: Authors

Smart growth boosts ecological integrity in both short- and long-term scales and promotes the quality of life based on the expanding urban transport options, creating jobs and increasing housing choices. The Urban Land Institute in the United State defines smart growth as development which is environmentally sensitive, economically viable, community based and sustainable (Urban Land Institute, 2003). Song (2005) believes that smart growth constitutes a set of planning, regulation and development methods where compact construction, infill development and adjustment of street and parking standards are used in order to make effective use of land (Song, 2005).

The review of the definitions and literature on smart growth guides us to a set of points in relation to features and dimensions of smart development and growth. Drawing on other definitions, this book explains and defines smart growth as follows:

> *Smart growth refers to a development that tries to revitalize the central and older parts of cities, support walking and cycling and preserve open and public spaces and natural lands. Smart growth controls desired development inside and on the edges of a city and follows a logical process to create communities for a better life. Smart growth seeks to achieve an integrated system of transport that supports compact developments and mixed uses in urban areas as opposed to car-oriented and fragmented developments in suburbs. Smart growth leads to accessible patterns of land use, improvement of transit opportunities, creation of healthy and viable communities and a decline in the costs of public services. Smart growth enables communities to grow in a way that can support economic and business development and create empowered neighbourhoods with a sustainable range of dwelling, trade and transit options. Smart growth provides solutions for problems of communities that have experienced highly dispersed growth in the past 50 years. In fact, this type of development places measures to promote civic life and social vitality, improve public transport and reduce the adverse environmental effects at the forefront of urban planning considerations. The revival of the city as a healthy and active environment that can secure a good future for all citizens is the main goal.*

Smart growth or smart planning for urban growth

Urban sprawl has caused extensive economic, social and environmental damage to human settlements. However, smart growth tries to deal with the consequences and negative effects of traditional urban development and provide a better model instead. A review of the literature shows that planning based on the smart growth approach cannot be confined to a few policies. Smart growth can be considered a programming style that tries to provide an appropriate response to the problems of fragmented urban development and prevent the scattered development of cities and the destruction of valuable natural spaces around them (Vanthillo & Verhetsel, 2012). Smart growth requires cooperation between both the public and private sectors in order to guide development towards the urban centres in cities and existing suburbs. However, some researchers believe that smart growth means different things for people with different aspirations (Yang, 2009). Next, some of the interpretations are briefly mentioned.

Smart growth vs. urban sprawl

Burchell (2000) states that smart growth is in contrast to urban sprawl and redirects urban development towards the central and inner parts of cities in an attempt to control urban development. In his view, smart growth occupies less land and natural resources, while still driving urban development (Burchell, 2000). Downs (2005) provides a similar definition. In his view smart growth is a logical response to the threats and problems arising from the continued traditional or scattered growth (Downs, 2005). Turner (2006) defines smart growth as planning to control confused and chaotic urban development and stresses the accessibility of housing, workplaces and leisure areas for all citizens (Turner, 2006).

Staley (2004), as a researcher and expert in the field of smart growth, describes the concept as follows:

> Smart Growth is the effort of the present era to tackle the illogical spread and development of cities. Smart Growth management is a means of exercising such control. Smart Growth does not seek to limit growth, but tries to help growth happen with proper respect to social and environmental issues.
>
> *(Staley, 2004)*

The smart growth movement prevents the negative impacts of urban sprawl and effectively uses natural and environmental structures to deal with scattered development. This type of development tries to reduce different types of economic inequalities and protect the natural environment (Theart, 2007).

Smart growth: an opportunity for communities

According to the Smart Growth Network (SGN) (2006), smart growth provides citizens with an empowered community, with various choices and considerable individual freedoms. This type of development seeks to protect public capital and achieve a dynamic and growing environment that can transfer a proud heritage to future generations. The smart growth method of construction challenges work and life. Such a development deems urban areas as not just places to live in but as a means to promote health and happiness and achieve prosperity and the desired life (Smart Growth Network, 2006).

Smart growth from the perspective of transit

Smart growth can also be deemed a planning theory (urban and regional) with a transit approach that stresses "preventing the fragmented spread of cities" and "growth in inner cities". This type of development supports mixed and compact uses and a public transit, pedestrian-oriented and cycling-friendly community. It constitutes planning to control the development of suburbs and focus on balanced development in the city centre that, in turn, prevents the illogical spread of access networks and increasing crowds in the suburbs (Handy, 2005). Figure 1.3 shows the importance and extensive use of bicycles in Amsterdam.

The Surface Transportation Policy Project and the Environmental Protection Agency define smart growth as a kind of compact urban development "with high walkability and access to public transport" that is considered a solution to controlling scattered suburbs (Surface Transportation Policy Project, 2003).

Smart growth supports a social response to urban issues and stresses the development of neighbourhood relations. To this end, its emphasis is more on walkable environments. Therefore, smart spatial planning uses medium urban density as a supportive means in order to achieve its goals (Environmental Protection Agency, 2013).

Smart growth for improving the quality of life

The theory of smart growth stresses compact urban design and planning to improve residents' quality of life and ultimately achieve sustainable development. In fact, the idea tries to align the development of neighbourhoods, cities and housing areas with the environment and communities and thereby increase the quality of life (Shapiro, 2005).

FIGURE 1.3 Bicycle and pedestrian lanes on a street in Amsterdam

Source: Borba, 2014

Smart growth: a solution for housing

This type of development means a rational and intelligent response to basic housing needs, which are created with population growth in mind and efforts to achieve economic prosperity and require the consensus of political, economic (the housing market) and urban planning efforts. Meanwhile, smart development, with an emphasis on planning mixed uses and denser housing construction, tries to protect open spaces and the sensitive environment surrounding the cities (Staley, 2004).

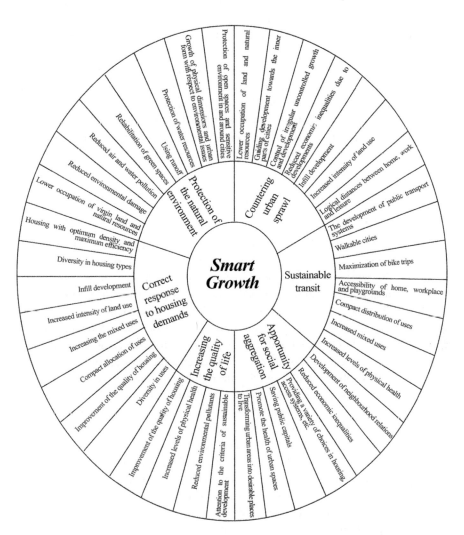

FIGURE 1.4   Components and pillars of smart growth and the relationship between them

Source: Authors

Nature conservation with smart growth

Smart growth creates a balance between the environment and development. It both advances urban development and protects open spaces and vulnerable nature and water resources (Bunce, 2004). Smart growth can be used as a strategy for addressing the continuing challenge of development towards green areas around cities (Hodges, 2009). It is an appropriate framework for cities, which have been faced with intense pressure from development and seek appropriate principles and policies for the restoration of their infrastructure and green spaces. Smart growth means land development with an emphasis on preserving the natural environment by minimizing dependence on cars, reducing air pollution and creating more efficient investment in infrastructure (Alexander & Tomalty, 2002). Smart growth includes a range of strategies for the development and protection of natural resources that help protect the natural environment and makes communities more attractive, economical, diverse and resilient. Smart growth constitutes the efforts of communities in the management and direction of growth in a way that minimizes damages to the environment, reduces scattered growth and makes cities liveable (Smart Growth Network, 2006). Figure 1.4 shows the components and pillars of smart growth and the relationship between them in detail.

Smart growth: an effective step towards urban sustainability

In 1972, a study entitled "Limits to Growth" was carried out by the Club of Rome[1] on the continuous effects of economic growth. In parts of this study, whose primary focus was the transition from exponential economic and population growth to a global balance, the realization of stable economic conditions alongside ecological stability in the distant future were discussed (Ogola, 2007). Many researchers consider the meeting of the Club of Rome a pioneer for sustainable development. The term sustainable development and its wide application emerged after the Brundtland Report was presented in 1992. With the increasing attention to the need for research on this subject and focus on survival, sustainable development became mandatory and went so far as to become the ultimate goal of urban projects. This goal changed the direction of most development (in different forms and contexts) towards sustainability (Drakakis-smith, 1995). Therefore, sustainable development was recognized as a global paradigm. (Bartelmus, 1999; Harris, 2003; Jensen, 2013; Royal Government of Bhutan, 2012; Schuftan, 2003). Sustainable development in its original form included three components: economic, social and environmental, but later on, three components of justice and equality, viability and tolerance (as the sum of previous components) were added to it. Therefore, it is conceptually close to the concept of smart growth and its goals.

Leitman (1993) believes that cities in developing countries need to increase the efficiency of the transportation infrastructure in order to survive in a globally competitive economy (Leitman, 1993). In fact, a large number of cities suffer from inefficient management in the field of urban transit systems and in some cases do not have the financial ability to maintain their existing transit systems (Bartley, 1996). However, many researchers have stressed the economic impacts of urban transit systems on economic conditions and highlighted the development of such systems as a critical factor in urban development (Davey, 1996; David-son & Payne, 2000; Peltenburg, Davidson, Wakely, & Teerlink, 1996). For example, Labi et al. (2008) pointed to the role of macro-transit systems in production and employment and stated that the manner of distribution and location of businesses (industrial, trade, home and institutional) is significantly affected by the planning of roads among cities. To them, a proper pattern of movement for goods and people can reduce demand for travel and can be used as a means for achieving sustainable growth at the macro (city) and micro wisdom (local communities) levels (Labi, Gkritza, Sinha, & Mannering, 2008).

Other researchers, such as Berke, Godschalk, Kaiser, and Rodrigerz (2006), also highlighted the influence of smart growth policies on sustainability of local and urban communities. Some experts also state that allocation of land to compact development as a method of sustainable land use planning can be used to "create equality for all citizens by increasing citizens' access to urban land uses and social mobility" (Berke et al., 2006). In a study on the social effects of urban sprawl, Stevenson (1995) points to the relationship between "urban poverty" and "lack of access". In his view, urban poverty is visible in poor public transit services. He studied transit networks in some cities in developing countries and found that these systems develop faster in streets located in high-income and wealthy areas, but is inefficient and inappropriate in low-income areas, which are isolated by the street systems (Stevenson, 1995).

Another effect of urban sprawl on sustainable development is the disruption of connections among local communities. The increase in urban routes leads to heavy traffic and unidirectional routs that increase walking distances. This, in turn, reduces cross-connection among urban areas, resulting in the separation of local communities and dividing them into several distinct sections (Goldstein, 1990; Tszmokawa & Hoban, 1997).

Another major issue that is affected by urban sprawl is environmental sustainability. Several studies have been carried out on the negative effects of urban sprawl on the urban environment. Such effects include noise and air pollution and climate change. Particulate pollutants and greenhouse gases produced by vehicles have a wide range of impact. This means that when they are emitted in the air, their concentration is maintained over a long distance from their source. Another environmental impact of fragmented urban development is the increased levels of energy consumption. Research shows that about 20 percent of the energy produced in the world is consumed for transportation, about 60 to 70 percent of which is used for transit. More than half of the oil extracted in the world is also used in the transportation sector, mainly within cities (Department of Infrastructure, 2012; Shafiea, Omara, & Karuppannan, 2013).

Given all this, it can be concluded that smart growth policies can prepare the ground for operationalization of a sustainable development paradigm. In fact, it seems that setting the macro concept of sustainability as a goal is not enough for modern cities and the development should progress in all parts of the city, in line with the principles of sustainable development. The effects of economic development on social, economic and environmental grounds should be examined in urban planning and design based on smart growth. This assessment continues to align the development project with values such as equality, viability and tolerance. From this perspective, smart growth is an efficient means to achieve sustainable urban development.

Goals and objectives of smart growth

Understanding the goals and objectives of smart growth requires an investigation of its origin and history. In 1996 and 1997 an urban plan and two books laid the building blocks for smart growth. Although none of these three documents and guidelines offer a definition of the concept, they do suggest what types of developments constitute smart growth and what goals are intended by this approach in urban life. Also, ten important organizations in the United States have done valuable research on a national scale in the field of smart growth, and each of them has described goals or strategies of smart development or growth.

In a book entitled *Sprawl and Politics: The Inside Story of Smart Growth in Maryland*, Frece (2008) wrote: "It was in Room 217 in the spring and summer of 1996 that Maryland's Smart Growth initiative was born in the Maryland State House". The term smart growth was first used by then Governor Parris N. Glendening in a project titled "Smart Growth and Neighbourhood Conservation Initiative" (Frece, 2008). The initiative referred to logical planning of land use and included the following eight principles (Lewis, 2011):

- Development is concentrated in suitable areas;
- Sensitive areas are protected;
- In rural areas, growth is directed to existing population centres and resource areas are protected;
- Stewardship of the Chesapeake Bay and the land is a universal ethic;
- Protect the natural environment by reducing consumption of resources;
- Formulation of appropriate economic mechanism for the four protection principles;
- Conservation of resources, including a reduction in resource consumption, is practiced;
- Funding mechanisms are addressed to achieve these visions.

In the United States, the first smart growth plan was developed at the state level based on regulating land use. This plan adjusted the horizontal growth of residential areas such as towns or villages and pursued protection of natural resources as one of its main objectives. Thus, urban development in this plan occurred in available population areas.

In 1997, the APA published the book *Policy Guide on Smart Growth* (American Planning Association, 1997). In the same year, the Natural Resources Defense Council published *The Toolkit for Smart Growth*, which presented policies for promoting compact growth, mixed use and transit-based development (Talen & Knaap, 2003) (Knaap & Talen, 2005). A review of these sources shows that the logic behind smart growth can be stated as follows (American Planning Association, 1997; Frece, 2008; Knaap & Talen, 2005):

- Whether its causes are economic forces, consumer preferences or misguided public policies, the dominant form of urban development over the post-war period can be characterized as urban sprawl;
- Urban sprawl can be defined as development that is low density, unplanned, automobile dependent, homogeneous and aesthetically displeasing;
- Urban sprawl has adverse effects on environmental quality, social cohesion, government finance and human health;
- Urban sprawl, and its associated evils, can be mitigated by policies that promote compact urban growth, mixed land uses, bicycle- and pedestrian-friendly environments, public transit, urban revitalization and farmland preservation.

According to Ferrier, Caves, and Calavita (2016) smart growth is a multi-scalar task that needs the collaboration of governance, citizens and the private sector in different scales from regional and beyond city boundaries to a local scale in the heart of a neighbourhood. They suggest six principles for smart growth in *The Challenges of Smart Growth: The San Diego Case* as follows:

- Creation of a more compact urban form by limiting sprawl at the metropolitan fringe through urban growth boundaries (UGBs) and open space conservation.
- Revitalization of existing communities through infill/densification and good community design while optimizing existing public facilities.
- Enhancement of the tax base of inner-city and first-ring suburbs through regional tax-base sharing. Also, the creation of affordable housing in suburban areas through regional fair-share housing.
- Redesign of old and new developments on the basis of New Urbanism principles that call for mixed use centres, job–housing balance, pedestrian-friendly communities, grid-street patterns, alleys, porches and other design elements that make neighbourhoods vital and diverse.
- Reorientation of the transportation system to reduce dependency on the automobile through land use measures as in the fourth bullet point, reallocation of funds of transit, and monetary disincentives, such as higher gasoline tax.
- Preservation of wildlife habitats, prime agricultural lands, and open space, especially at the urban fringe.

None of these sources (as the first specialized and professional texts on smart growth) provide a clear definition of the concept. However, they present different

objectives for smart growth – ranging from optimal locations to public sector financial management. In recent decades, many international institutions have defined
smart growth. However, many scientific resources still define the concept through
its representations or objectives. In the rest of this chapter, the objectives and goals
of smart growth are discussed from the perspective of some professional bodies and
academic institutions.

The U.S. Environmental Protection Agency (EPA) (2002), as the country's most
important authority in environmental issues, defines smart growth as a "development that serves the economy, the community, and the environment. It changes
the terms of the development debate away from the traditional growth/no growth
question to how and where should new development be accommodated". The
EPA states the following four objectives for smart growth) U.S. Environmental
Protection Agency, 2002):

- **Healthy communities** – that provide families with a clean environment.
  Smart growth balances development and environmental protection, accom-
  modating growth while preserving open space and critical habitat, reusing land
  and protecting water supplies and air quality.
- **Economic development and jobs** – that create business opportunities and
  improve the local tax base, that provide neighbourhood services and amenities
  and that create economically competitive communities.
- **Strong neighbourhoods** – that provide a range of housing options, giving
  people the opportunity to choose the housing that best suits them. They maintain and enhance the value of existing neighbourhoods and create a sense of
  community.
- **Transportation choices** – that give people the option to walk, ride a bike,
  take transit or drive.

From the perspective of the U.S. Department of Housing and Urban Development
(2003), one of the aims of renovation of local communities is to improve the quality
of life and sustainable economic development through implementing smart growth
strategies on a regional scale. The department defines smart growth as an attempt
to achieve three objectives as follows (U.S. Department of Housing and Urban
Development, 2003):

- **Increasing housing options** – change the stereotypical traditional single-
  family residential housing types and expand housing options such as multi-
  family, multi-unit housing.
- **Integrating land uses with housing** – integrate land uses to re-create urban
  and suburban neighbourhoods that integrate residential, commercial and recreational functions, thus reducing the heavy dependence on automobiles.
- **Elevating design** – vertical elevating design is a key to make urban areas more
  compact, more mixed and denser. The design involves more than physical

appearance; it includes designing infrastructure, recreation and transportation systems and, more broadly, land use systems to create attractive areas that create a sense of place.

The U.S. Department of Agriculture (USDA) (2001) defines smart growth as a comprehensive approach that presents a set of land use policies affecting the pattern and density of new developments and developments in urban and suburban areas. These policies do not prohibit development outside the areas mentioned and are not a threat to individual property rights. The goals and objectives of smart growth are defined by the USDA as follows (U.S. Department of Agriculture, 2001):

- Locating new development in centre cities and older suburbs rather in fringe areas;
- Supporting mass transit and pedestrian-friendly development;
- Encouraging mixed use development (e.g. housing, retail, industrial);
- Preserving farmland, open space and environmental resources.

The APA defines smart growth as a development that leads to the following characteristics (American Planning Association, 2002):

- Have a unique sense of community and place;
- Preserve and enhance valuable natural and cultural resources;
- Equitably distribute the costs and benefits of development;
- Expand the range of transportation, employment and housing choices in a fiscally responsible manner;
- Value long-range, regional considerations of sustainability over short-term incremental geographically isolated actions;
- Promote public health and healthy communities.

The SGN defines smart growth as the opposite of traditional urban growth and believes that this type of development is more than paying attention to urban centres. The main objectives of smart growth according to the SGN are as follows (Smart Growth Network, 2006):

- **Neighbourhood liveability** – neighbourhoods should be safe, convenient, attractive and affordable;
- **Better access, less traffic** – emphasizing mixing land uses; clustering development; and providing multiple transportation choices to manage congestion, pollute less and save energy.
- **Thriving cities, suburbs and towns** – guiding development to already built-up areas to save investments in transportation, schools, libraries and other public services and to preserve attractive buildings, historic districts and culture landmarks.

- **Shared benefits** – eliminating divisions by income and race and enabling all residents to be beneficiaries of prosperity.
- **Lower costs and lower taxes** – taking advantage of existing infrastructure, relying less on driving and saving investment for other things.
- **Keeping open space**.

The Sierra Club states the following objectives for smart growth (Parfrey, 2003):

- Liveable communities, designed for people rather than for automobiles;
- Closeness to nature and permanent conservation of important lands;
- Viable public transit at the city and metropolitan area level is needed to support compact forms of development;
- Revitalization of older suburbs and downtowns and rundown commercial areas;
- Urban growth boundaries;
- Long-term visions for communities and regions.

The Trust of Public Land (TPL) is a national non-profit organization whose main purpose is to protect land from development pressure and provide public access to public lands. TPL does not inhibit the growth of human settlements, but believes that the typical development pattern based on sprawl can harm the environment, urban form and communities. Activists in TPL believe that the most important point in achieving smart growth is to help more people believe that life on smaller areas of land can have higher vitality. According to TPL, growth will be really smart if human society decides to preserve lands for leisure, for protecting natural resource, and as open spaces in the settlements. Such decisions turn developments into smart growth, out of which intensive development occurs. The following are the goals of smart growth according to the TPL (Trust for Public Land, 1999):

- Conserve lands to protect for recreation, community character, conservation of natural resources and open space;
- Make existing communities attractive and liveable enough to steer growth away from the countryside;
- Identify clearly both "desired development zones" and the lands it wants to protect;
- Ensure that existing neighbourhoods, including those with lower-income housing, will have full access to the system of parks and greenways.

The Home Builders Association defines smart growth as a political agenda, based on which local comprehensive plans are created and various housing choices are presented via land use planning. Also, smart growth provides the infrastructure necessary for new residential, commercial and industrial uses before developments and protects open spaces and the environment. The association presents six principles for effective smart growth as follows (Home Builders Association, 2002):

- **Meeting the nation's housing needs** – plan for the anticipated growth in economic activity, population and housing demand, as well as ongoing changes in demographics and lifestyles;
- **Providing a wide range of housing choices** – plan for growth that allows for a wide range of housing types to suit the needs and income levels of a community's diverse population;
- **A comprehensive process for planning growth** – identify land to be made available for residential, commercial, recreational and industrial uses and meaningful open space;
- **Planning and funding infrastructure improvements** – encourage local communities to adopt a balanced and reliable means to finance and pay for the construction and expansion of public facilities and infrastructures;
- **Using land more efficiently** – support higher-density development and innovative land use policies to encourage mixed-use and pedestrian-friendly development;
- **Revitalizing older suburban and inner-city markets.**

The Urban Land Institute as a national prestigious institution in the United States. It defines smart growth as development that is environmentally sensitive, economically viable, community based and sustainable. What to do and how it should be done are not detailed in smart growth, as it only provides a general direction for progress. Smart growth helps local communities recognize what kind of growth best meets their needs. The institute classifies the objectives of smart growth as follows (Urban Land Institute, 2003):

- Development is economically viable and preserves open spaces and natural resources;
- Land use planning is comprehensive, integrated and regional;
- Public, private and non-profit sectors collaborate on growth and development issues to achieve mutually beneficial outcomes;
- Certainty and predictability are inherent to the development process;
- Infrastructure is maintained and enhanced to serve existing and new residents;
- Redevelopment of infill housing, brownfield sites and obsolete buildings is actively pursued;
- Urban centres and neighbourhoods are integral components of a healthy regional economy;
- Compact suburban development is integrated into existing commercial areas, new town centres and/or near existing or planned transportation facilities;
- Development on the urban fringe integrates a mix of land uses, preserves open space, is fiscally responsible and provides transportation options.

In order to examine the objectives of smart growth, the principles presented in the resources provided in this chapter are presented in Table 1.5 under eight dimensions: land use planning, infrastructure planning, physical planning, transit planning, environmental planning, economic planning, housing planning and social planning.

TABLE 1.5 Objectives of smart growth based on the main dimensions

Dimensions	Objectives of Smart Growth
Land use planning	Comprehensive, integrated and regional land use planning
	Development based on optimum location
	Creating and integrating diverse uses
	Encouraging mixed developments
	Injecting required uses to existing neighbourhoods
Infrastructure planning	Increasing the capacity of existing infrastructure services
	Maximizing the utilization of existing equipment and infrastructure
Physical planning	Compact physical development
	Vertical design
	Infill development
	Rehabilitation of inner-city areas (worn out and old neighbourhoods and centres)
	Rehabilitation of existing suburbs
	Creating more attractive places
	Protection of valuable monuments, historical places and cultural heritage
	Increasing spatial equity
Transit planning	Development of public transit network
	Pedestrian-based development
	Bicycle-friendly environment
	Expanding the range of transit options
	Increasing accessibility in urban areas
	Reducing traffic
Environmental planning	Protecting open spaces, farmland and sensitive areas in the environment
	Reducing energy consumption
	Reducing the distance between urban and natural areas
	Creating healthy communities (public health promotion)
	Creating a balance between development and natural areas in need of protection
	Protecting certain animal and plant species
	Protecting environmental resources (water and air)
	Directing and using surface runoff
Economic planning	Fair distribution of the costs and benefits of development
	Reducing costs and taxes
	Creating societies with a more competitive economy
	Creating different job opportunities
Housing planning	Increasing housing options for different households
	Meeting housing needs appropriately
	Building housing in existing urban areas
	Public participation in the construction of housing
	Identifying and strengthening unique features of communities in housing construction
Social planning	Creating socially sustainable neighbourhoods (creating a sense of social solidarity)
	Increasing the sense of place
	Enhancing the quality of life
	Increasing social vitality

Source: Adapted from: (American Planning Association, 2002; Knaap & Talen, 2005; Litman, 2015b; Smart Growth Network, 2006)

In search of the principles of smart growth

As mentioned before, smart growth seeks to achieve compactness in land use patterns and rely on walking and cycling in the urban transit system. Smart growth and the New Urbanism rely on it to create streets with more connections than that of old access networks. Routes should be responsive and reduce car traffic. On the other hand, smart growth deems social interaction necessary for the development of neighbourhood relations, and to this end, emphasizes creating walkable environments more. Azizi (2004) points out that reducing physical distances in compact cities as one of the strategies of smart growth can reduce urban traffic and the air pollution resulting from car movements. The efficient use of urban land protects agricultural land around cities against urban development. He also believes that smart growth includes principles that lead to equality and increases the capabilities and diversity of transit choices that is ultimately to the benefit of low-income households (Azizi, 2004).

Most details of smart growth emerged from the meeting of some urban planners and architects in 1991 at the invitation of the Local Government Commission, led by Peter Katz in Yosemite National Park's Ahwahnee Hotel in California. This group decided to determine a principled structure to resolve issues related to urban sprawl. The introduction to their statement reads (Calthorpe et al., 1991):

> Existing patterns of urban and suburban development seriously impair our quality of life. The symptoms are: more congestion and air pollution resulting from our increased dependence on automobiles, the loss of precious open space, the need for costly improvements to roads and public services, the inequitable distribution of economic resources, and the loss of a sense of community. By drawing upon the best from the past and the present, we can plan communities that will more successfully serve the needs of those who live and work within them. Such planning should adhere to certain fundamental principles.

According to the Local Government Commission, achieving optimal communities for a good life requires attention to the principles presented in Tables 1.6 and 1.7 at local and regional levels (Calthorpe et al., 1991).

According to the Local Government Commission (1991), achievement of these principles requires an updated general plan that includes these principles. The plan must be developed through an open and collaborative process and should explain all proposed plans together with visual models (Calthorpe et al., 1991).

Porter (2002) believes that to have a correct understanding of smart development, the basic principles must be addressed first. These principles are a comprehensive set of accepted inter-related ideas about the desired form and social character (Porter, 2002). The framework is often based on the long-term yet comprehensive

TABLE 1.6 Local principles needed to achieve optimal communities

- All planning should be in the form of complete and integrated communities containing housing, shops, workplaces, schools, parks and civic facilities essential to the daily life of the residents.
- Community size should be designed so that housing, jobs, daily needs and other activities are within easy walking distance of each other.
- As many activities as possible should be located within easy walking distance of transit stops.
- A community should contain a diversity of housing types to enable citizens from a wide range of economic levels and age groups to live within its boundaries.
- Businesses within the community should provide a range of job types for the community's residents.
- The location and character of the community should be consistent with a larger transit network.
- The community should have a centre that combines commercial, civic, cultural and recreational uses.
- The community should contain an ample supply of specialized open space in the form of squares, greens and parks whose frequent use is encouraged through placement and design.
- Public spaces should be designed to encourage the attention and presence of people at all hours of the day and night.
- Each community or cluster of communities should have a well-defined edge, such as agricultural greenbelts or wildlife corridors, permanently protected from development.
- Streets, pedestrian paths and bike paths should contribute to a system of fully connected and interesting routes to all destinations. Their design should encourage pedestrian and bicycle use by being small and spatially defined by buildings, trees and lighting, and by discouraging high-speed traffic.
- Wherever possible, the natural terrain, drainage and vegetation of the community should be preserved with superior examples contained within parks or greenbelts.
- The community design should help conserve resources and minimize waste.
- Communities should provide for the efficient use of water through the use of natural drainage, drought-tolerant landscaping and recycling.
- The street orientation, the placement of buildings and the use of shading should contribute to the energy efficiency of the community.

Source: Adapted from: (Calthorpe et al., 1991)

TABLE 1.7 Regional principles needed to achieve optimal communities

- The regional land use planning structure should be integrated within a larger transportation network built around transit rather than freeways.
- Regions should be bounded by and provide a continuous system of greenbelt/wildlife corridors to be determined by natural conditions.
- Regional institutions and services (government, stadiums, museums, etc.) should be located in the urban core
- The materials and methods of construction should be specific to the region, exhibiting a continuity of history and culture and compatibility with the climate to encourage the development of a local character and community identity.

Source: Adapted from: (Calthorpe et al., 1991)

prospects for communities and can help determine the correct direction of social progress. Any society can use specific methods and programs with respect to their particular circumstances and regional needs in order to implement smart growth principles (Porter, 2002, p. 7).

In this regard, the SGN, as the most active organization in the field, adopted the Ahwahnee Principles to develop the basic principles of smart development in ten sections (Smart Growth Network, 2002):

- Mix land uses;
- Take advantage of compact building design;
- Create a range of housing opportunities and choices;
- Create walkable neighbourhoods;
- Foster distinctive, attractive communities with a strong sense of place;
- Preserve open space, farmland, natural beauty and critical environmental areas;
- Strengthen and direct development towards existing communities;
- Provide a variety of transportation choices;
- Make development decisions predictable, fair and cost-effective;
- Encourage community and stakeholder collaboration in development decisions.

Next, each of these principles will be discussed in detail.

The first principle: mixed use

Although separated uses were initially used with the intention to protect neighbourhoods against pollution from urban industries and activities, this approach finally resulted in a pattern of land development that increased the distance among shopping centres, residential neighbourhoods and schools so that people had to use car to access them (Gyourko & Rybczynski, 2000). In fact, the regime governing the expansion of suburbs lead to the separation of uses. The current zoning limitations based on the promotion of low density and spread of single uses aimed at maintaining a distance between residential uses and industrial gases, congestion, crowding and other problems caused by the mixing of uses, all can cause a type of development that requires access to extensive public transit and, in the absence of public transit, leads to greater use of private cars.

In contrast to the view that mixing uses leads to hustle and bustle and other problems in cities, Jacobs (1961) believes that "[i]ntricate minglings of different uses in cities are not a form of chaos. On the contrary, they represent a complex and highly developed form of order". Based on this approach, mixing various commercial, residential, recreational, educational and other uses is one of the most important ways of promoting social vitality and sense of place (HosseinZadeh Dalir & Hoshyar, 2005). The APA also points out that mixed uses can meet the needs of community residents and play an effective role in attempts to enliven the urban areas (American Planning Association, 2006).

In fact, when different uses with different scales come together, more people go to the streets, thereby increasing the vitality of space. On the other hand, because of shorter paths of access to different uses, more people are attracted to the area, making the compression of urban contexts possible. The concentration of a range of uses in an area is also cost-effective; therefore, it can be concluded that mixed uses can help economic flourishing in various fields. Accordingly, this type of development has many advantages and can result in more vitality, sustainability, sociability, proper accessibility, enhanced level of security and increased social interactions and can lead to more effective use of the existing infrastructure.

According to what was mentioned earlier, utilization of mixed uses has become one of the most important features of smart growth, and theoretically, the use of complex and diverse uses is considered one of the easiest ways to promote "diverse housing and transit options in the vicinity of workplaces, shops and schools" (Dong, 2010). From a smart growth perspective, in an ideal society, residents at the local level can do activities such as working, shopping, receiving services, enjoying leisure and sending children to school, and each of these activities can take place via different options (Yang, 2009).

From among the principles of smart growth, the two principles of mixed uses and compact design of buildings focus directly on the issue of land use. In addition, some other principles of smart growth – such as creating walkable communities and providing a variety of transit options – can influence land use patterns. In fact, a successful model of land use results from a joint effort between different principles, taking into account all aspects of the physical environment (Yang, 2009, p. 88). Figure 1.5 illustrates a pedestrian zone with mixed use regulation in Istanbul.

FIGURE 1.5 Mixed-use pedestrian zone in Istanbul, Turkey

Source: Authors

a) **Benefits of mixed use**

The rules of smart growth illustrate the importance of the development of mixed uses in new and existing areas of a city. Many scientific studies on smart growth stress the need for a new perspective to move away from single uses towards integration of various uses in urban land and buildings. According to this view, in mixed use circumstances, residents will be less dependent on cars because of their proximity to workplaces, schools and recreation areas. Decreased traffic and driving time can reduce traffic congestion and air pollution. Also, most residents use sidewalks and other spaces in their environment, which in turn increases their interaction with each other and improves the social security. Economically, proponents of smart growth believe that the establishment of business centres in residential areas leads to increased land values because residents will recognize the value of a shorter commuting distance between the workplace and residential areas (Reese, 2011). Such an approach can consequently help to increase local taxes.

Shops and business centres also benefit from areas with mixed uses that have the ability to attract people. With a large number of people present to shop, economic activities will flourish. Uses near residential areas can do better in attracting people during the entire day (Hosseinzadeh & Safari, 2012). If a home is located in walking distance from shopping or employment centres, more people can access them on foot or by bicycle. In addition, mixed uses can attract a more diverse population and create powerful business centres, which in turn, can support a public transit system more effectively. Thus, the location of shops in the local shopping areas and city centres will reduce car trips to malls and stores on the outskirts of cities. Stores with better access increase walking, cycling and use of public transport, and unlike large commercial stores in suburbs, can reduce car trips and commuting distances. It should also be noted that local shopping and urban centres lead to single-park trips (park in one place and then walk to several stores instead of driving from one shop to another) that reduce the total demand for parking (Abley & Turner, 2011).

Mixed uses provide the possibility of shopping, relaxing and doing all activities people need in a walkable distance from residential areas (especially for those who are not fortunate enough to buy a car or cannot drive due to their age) and allow all people to have a decent and enjoyable life. Also, mixed uses reduce the negative impact of traffic on the roads, and the capacity of the road transport can satisfy the needs of people for a longer period (Theart, 2007).

b) **Ways to measure mixed use**

Almost every scale of smart growth measures mixed use in one way or another. Robert Cervero and Kockelman (1997) used two specific variables to measure land use diversity. The first variable is vertical mixing, which is measured by the extent to which more than one use can be found in single business/retail units. The other is mixing activity centres, which is measured by the extent to which activity centres have a variety of uses in them. Miles and Song (2009) do not use the ratio of different uses for the measurement of mixed uses, but rely on the actual area of commercial, industrial and public lands and the number of stores on a block. Fleissig and Jacobsen (2002) believe that in a project, mixed use should be based on the number of different uses available in the area (Hagerty, 2012). Downs (2005) believes that among the policies of smart growth, mixed uses is the most likely to be implemented, because there are no serious objections against it (Dong, 2010).

c)   **Ways to achieve and implement more mixed uses**

One way to implement mixed uses is to describe them in the framework of zoning or other plans in urban planning. The SGN provides other innovative policies and strategies (Smart Growth Network, 2006), including financial incentives; changes in planning regulations or rules; comprehensive and subarea plans to identify the community goals; improved zoning techniques (such as flexible zoning, overlay zoning and impact zoning); grants on a regional planning scale; redevelopment of single use areas; adjustment of new uses; and reuse of closed, outdated or obsolete administrative-organizational uses to achieve the successful implementation of mixed uses (Theart, 2007, p. 8).

Mixed uses can be implemented through the establishment of spaces and amenities for daily use in urban neighbourhoods. The literature on smart growth suggests that it can be done through changes in the structure of urban zoning rules to specify the type of structure instead of the type of use. Under this idea, several uses of a building are available for developers and residents, and only the physical and aesthetic properties of recreational spaces are shaped by city managers. Another action that may encourage mixed uses is the use of state and federal funds to help public service workers live near their workplace. Parallel plans to upgrade the existing planning structures by urban managers and innovative zoning measures can also provide additional flexibility in development. The general purpose is to encourage flexibility in decision-making processes for development and enable people to live and work with a reasonable commuting distance (Reese, 2011, p. 20).

The second principle: compact building design

Today, the use of land has increased for residential, trade, recreational and other purposes. Nonpremeditated development patterns added to many suburban areas are developing and advancing. These patterns are generally in contrast with the needs and interests of future urban development patterns. In such circumstances, smart growth encourages communities to follow a compact development model and try to correct spatial planning to minimize the impact of new buildings on undeveloped areas.

A compact city is in a form suitable for walking, cycling and public transit along with a density which can encourage social interaction. The main characteristic of such a city is the density and composition of the land use that can reduce up to 70 percent of the distances travelled in the city. Tsai (2005) believes that distinguishing different patterns of compactness and sprawl is possible by looking at variables such as higher land use and land consumption per capita and improper distribution of uses in sprawling cities versus increased compactness and interconnection in compact cities (Tsai, 2005, p. 43).

There are three perspectives on the issue of increasing density in compact cities: increasing density in the entire urban area, increasing the density of city blocks and increasing density in one urban core instead of in multiple cores[2] (Gordon & Richardson, 1997). However, dense buildings are more important in achieving desired urban growth, because they provide better opportunities for growth and development and make efficient use possible.

Quoting Cervero and Kockelman (1997), Hagerty writes: "Density is the necessary element that allows the other principles [of mart growth] to occur. In order to achieve a self-sustaining mix of commercial and residential, a minimum residential density is required". This density allows the creation of a neighbourhood with mixed uses that encourage walking, as well as supports the retailers trading on the ground level. In fact, smart growth is more than just density, but density is still a key principle underlying it (Hagerty, 2012).

Theart also quoted Kackar and Preuss (2003), stating that

> higher densities also reduce the impact of the built area on the environment which ensure that people are accommodated on a smaller geographic area which lead to the fact that less land be consumed and that valuable open spaces, farmland and ecologically sensitive areas be protected.
>
> *(Theart, 2007, p. 9)*

In fact, a compact design reduces the exertion of the pressure of development on green spaces, increases water permeability and improves the productivity of urban resources (Reese, 2011, p. 21).

The SGN (2002) pointed out in this regard that smart growth makes it possible for communities to create compact buildings to thwart poor and scattered development of lands. Under this pattern, cities must be designed in a way that creates more open space and allows more efficient use of land and resources. By encouraging buildings to grow more vertically than horizontally and correcting the combination of buildings with parks and open spaces, communities can reduce new construction and maintain more green space.

In fact, higher density in existing areas can help us protect open spaces in metropolitan areas and accommodate an acceptable level of urban population in these areas via a more effective use of capacity in residential spaces (Hagerty, 2012, p. 25). Thus, patterns based on higher density and compact buildings are more efficient and economical in terms of energy consumption, because they can reduce the amount of land needed and preserve it for future generations. One on hand, the need for land for urban development is reduced and, on the other, open undeveloped spaces are preserved. Undeveloped areas that can absorb and purify rainwater reduce damages caused by storms and the need for water drainage and decrease the number of contaminants entering watercourses, rivers and lakes.

On the other hand, local officials have found that in compact neighbourhoods, providing per capita services such as water, sewerage collection, electricity, telephone and other facilities can be done with lower costs and higher ease compared to scattered ones (Rahnama & Abaszadeh, 2006, p. 48). In fact, the need for water, sewage and gas pipes, as well as electricity and telephone wires in denser neighbourhoods is lower than it is in scattered neighbourhoods. As a result, the provision and maintenance of services for denser communities is easier for local government. In the book by Kackar and Preuss (2003) entitled *Creating Great Neighborhoods: Density in Your Community*, the Office of Technology Assessment of the United States reported that "it cost a western city $10,000 more to provide infrastructure to a lower density suburban development than to a more compact [city]". The Urban Land Institute found that the cost of building infrastructure per each housing unit reduces dramatically with increasing density. Accordingly, it is clear that creating commercial and residential uses in higher-density areas reduces the costs of development and increases tax revenue for local governments without imposing an oppressive financial burden on property owners (Theart, 2007, p. 9).

The compact and dense contexts provide a variety of housing options in different sizes and prices (in the form of duplex townhouses, flats, and single-family detached housing) in walkable distance from the shops, offices, recreation areas and public transit system. This helps us ensure that a wider range of individuals and groups in society are able to live in a residential area. Moreover, such density supports the economy and the transit system and creates greater social unity in all residential neighbourhoods (Theart, 2007, p. 9).

The California experience demonstrates that creating a denser community that doubles household density reduces the movement of private vehicles by 20 to 30 percent, because people have access to easier and less expensive transit than private vehicles (Cohen, 2018).

Achieving smart growth in its real sense requires higher density that improves public transit efficiency. In this regard, Reese (2011) believes that a density of at least six to eight households per four hectares (an acre) is essential to support rail transit. Some of the problems are related to parking requirements and lack of awareness about the benefits of intensive development. For the development of new housing in residential neighbourhoods, smart growth encourages older styles of local development, including constructing houses near the front part of the land lots to help better define the street, using the element of porches in front of each house to increase opportunities for social interaction, placing parking behind the house and at the same time encouraging some to park their cars on the streets to slow the traffic and providing easier access to public green spaces. Based on smart growth policies, tax incentives and rewards should be provided to residents who attempt to carry out such measures (Reese, 2011, p. 21).

Transit-based regional planning can change the zoning system for lands located in the vicinity of the public transit system, with the aim of promoting mixed uses and higher densities and helping developers (Nahlik, 2014).

Intensive residential areas with higher density and mixed use can contribute more to reducing negative environmental impacts due to car use (Banister, Watson, & Wood, 1997). Indeed, despite the different opinions presented in the field, most theoretical texts stress the fact that a compact city can yield a sustainable urban form. Also, many studies suggest that residents of neighbourhoods with high density and mixed use are more likely to walk and have fewer inter-city trips than do residents of low-density neighbourhoods (Cervero, 2002). Therefore, we can say that the components of the urban physical form, including congestion, variety, connectivity networks and pedestrian environments, can affect the choice of vehicle. Given the relationship between land use and transit, reducing congestion and increasing distribution of service centres can change the pattern of transit and car trips (Litman, 2005).

So, we can say that smart growth leads to a development with an adequate level of use density that reduces the use of personal vehicles (Girardet, 2004). In this context, Marshall (2010) states that the city and inner-city layout contributes to the choice of vehicles and to the travel distance, and components such as population density, land use and public transit are associated with per capita travel (Hankey & Marshall, 2010). In an article titled "Smart Growth and Sustainable Development: Challenges, Solutions and Policy Directions" Alexander and Tomalty (2002) used 13 indicators to investigate the relationship between density and urban development in 26 municipalities in British Columbia, Canada. They found that density was associated with reduced use of cars, infrastructure efficiency and ecological and economic performance (Alexander & Tomalty, 2002).

Neighbourhoods can create denser plans. For example, they can construct higher buildings rather than wide ones along with surface parking. However, we should notice that the removal of barriers in existing zoning in line with the realization of compact design is difficult. For example, in many cases, the minimum requirements of plots in such zonings prevent the creation of more compact contexts. Also, many existing zoning regulations encourage neighbourhoods to build big and expensive houses that have higher tax benefits for the municipalities.

The third principle: creating a range of housing opportunities and choices

A significant portion of new construction and development in cities is related to housing. People have different housing needs and demands. Some single-person households prefer renting small flats. Young couples need smaller housing, while other families require dwellings with a larger number of rooms. Single middle-aged parents need to be closer to services, and the elderly seek local care. Yet most citizens should be able to continue to live near their friends and relatives despite their changing needs and stages of their lives. Social professionals such as police, firefighters and teachers should be able to find affordable housing in the vicinity of their workplace. Hence, neighbourhoods should provide a diverse range of options to their residents, including single-family houses of different sizes, houses with yards and gardens, shared houses, affordable housing for low-income households and customized houses for the elderly (USEPA, 2011).

In fact, changes in the demographic structure of families (based on census results) have led to heightened needs for the development of existing settlements. Hossein-Zadeh Dalir and Hoshyar (2005) quoted Porter (2002), stating that

> single-family detached housing designed for a family constituting parents and two children cannot satisfy the needs of single adults. However, adults with two salaries and without children, single parents, and elderly citizens constitute an important part of applicants for the housing market.

Add to this the limited housing opportunities for low-income people, who live in isolated communities with limited resources (Hosseinzadeh & Safari, 2012).

Studies on big and small cities show that most urban centres have a combination of a variety of housing options (such as apartments, single-family detached housing units, etc.), and the more we get away from the city centre, the more we are faced with large areas allocated to residential villas. This leads to a situation where each neighbourhood is formed by only a certain kind of dwelling, which is not affordable for many people. Although these housing types are located near the workplaces, schools or other specified activities in the lives of many individuals, in physical terms, they cannot respond to the needs of many households. For the principles of smart growth to be effective, an innovative plan is needed to create a combination of detached houses, flats and villas in the suburban areas. Although developments are normally determined based on free-market demands or social principles identified in planning documents, sometimes the residents undermine them based on their individual interests, without consideration of the consequences for the city as a whole (Theart, 2007, p. 10).

On the other hand, construction of new housing can also be a stimulus for business centres that are currently active during the day but suffer from the lack of pedestrian traffic and lack of users in the evening or on the weekends. Supplying a range of housing options to all households allows them to find their place in the smart growth community – whether it is a garden apartment or a traditional house in the suburbs – and thus adapt to the conditions of the time (Hagerty, 2012).

Based on all this, we can say that providing quality housing for people of all income levels is an integral part of smart growth strategy, because it can improve the households' quality of life and decrease the environmental costs of car-based development by supporting public transit systems and constructing commercial centres and other facilities in the vicinity of neighbourhoods. Under this principle of smart development, a society should be able to create the necessary infrastructure for the provision of quality housing for all income groups and respond effectively to the needs of different families. In fact, a combination of single-family and multi-family structures in existing neighbourhoods can reduce the severity of poverty, increase density, facilitate effective site selection for residences and employment alike, better satisfy the needs and requirements of diverse inhabitants, promote fairness[3] in utilization of network access and a public transit system and bring about many other benefits for the heterogeneous population of neighbourhood units.

The fourth principle: creating walkable neighbourhoods

Until the mid-1900s, urban communities and neighbourhoods were developed on pedestrian-level scale. In fact, they were designed to lead people to their destinations. However, in the past 50 years, scattered growth patterns and separation of uses have increased reliance on private cars and removed many of the characteristics supporting walkable communities. Traditional land use planning typically denied mixed uses, leading to longer journeys and reduced walking opportunities as a form of public transit options. This wrong approach against the development of mixed uses was exacerbated by private-sector fiscal policies that deemed mixed use as a threat to single-use approaches (Smart Growth Network, 2002).

Today, traffic engineers plan fewer sidewalks so that many new streets have no sidewalk or have only a footpath on one side. Based on car-oriented development, engineers and developers believe that sidewalks do not by themselves lead to walking. In fact, they are somehow right, because walking requires features such as the right combination of densities, different uses, connections, street intersections and neighbourhoods (Smart Growth Network, 2002).

However, this view has changed to a great extent in recent years. Studies suggest that walkable neighbourhoods improve pedestrian activities and social interaction between users and increase the sense of security in neighbourhoods by adding to the number of pedestrians. In fact, the purpose of the implementation of this smart growth principle is to place travel destinations, spaces, amenities and recreational facilities near residential areas so that residents can enjoy safe and pleasant walking for everyday activities (Theart, 2007, p. 11). In addition, locating services in a walkable distance with easy access for pedestrians reduces their dependence on cars, which in turn, improves physical health and air quality and eliminates the need for large parking spaces. Such neighbourhoods are cost-effective and help improve the social interactions of people with a safer and stronger sense of place.

Based on this principle, neighbourhoods can be constructed in a way that improves access for about one-third of the urban population, including the elderly, children or the very poor, and brings about many benefits for the environment.[4] For example, the correct design of sidewalks can reduce the need for driving and thus improve air quality. In sum, many amendments proposed to support network access has led to reduced levels of impenetrable grounds on passages and thereby reduced the flow of stormwater. Such strategies have many economic benefits for the city as well and improve air and water quality and reduce costs associated with them (Handy, 2005). In addition, these benefits encourage and motivate more neighbourhoods to adopt such a model and cooperate with public and private sectors in the development of walkable communities.

Many communities, especially those who are dispersed and highly car dependent, use a type of design that reduces pedestrian activity. Wide streets that lack the necessary infrastructure for pedestrians such as pleasant sidewalks, with a limited number of places for people to cross through, long blocks, etc. greatly reduces people's incentives for walking. Common residential designs often work as a deterrent for pedestrian activity (Smart Growth Network, 2002). Requirements such as a lot of building setbacks, minimum area for separation of plots and indirect pedestrian passages that create deadlocks increase the distance between urban travel destinations. Such restrictive factors are also seen in commercial designs. For example, many office buildings, hotels and other commercial uses are surrounded by vast parking lots and force pedestrians to pass through an ocean of parked cars (Ewing, 1999).

These barriers show how land use and design of neighbourhoods and urban areas play a central role in encouraging pedestrian-oriented environments. Neighbourhoods can increase walkable areas by creating multiple destinations and ensuring the balance of all types of movements through proper sizes of sidewalks and other passages (Smart Growth Network, 2002).

To achieve walkable neighbourhoods, some factors seem essential in urban design. For example, designing narrower streets, providing parking on the streets and unifying passages through parking lots help to slow down traffic and facilitate the use of space in the neighbourhoods. In vibrant commercial areas, buffer spaces between sidewalks and crosswalks and vehicle traffic on busy intersections make pedestrians feel more secure. Design standards determine the minimum width of the sidewalks relative to adjacent buildings and, while observing human–scale criteria, create intimate spaces for users of urban spaces. The principles and policies that can increase walkability in urban areas are discussed in detail in Chapter 5.

The fifth principle: fostering distinctive, attractive communities with a strong sense of place

One of the important principles of smart growth is the creation of societies with different cultures and architectural, aesthetic and social values. Smart growth tries to give identity to a town, region or neighbourhood so that the boundaries of communities are defined and a variety of physical spaces with special civil identities are formed so that people have a sense of place in their community and try to further promote its quality.

Contemporary development patterns have helped with the creation of large shopping centres and extended dwellings on the suburbs that – with the exception of slight differences in decoration – are not distinguishable from each other. This approach initially leads to lower costs, making construction profitable for some people, but it does almost nothing to maintain civil dignity or create a strong sense of association through which communities can earn their identities and identify with their neighbourhoods (Smart Growth Network, 2002).

By contrast, according to smart growth principles, not only should developments respond to the basic business or housing needs, but they should also help create unique and distinct communities. Smart growth seeks to cultivate a variety of physical environments with a sense of social pride, civic pride and an interrelated social fabric. This helps us achieve economic interests and high-quality communities with architectural and natural elements that reflect the wishes of all residents and in the long term preserve the vitality and economic value of local communities (Bohl & Schwanke, 2002; Lucy, 2002).

In fact, smart growth encourages communities that specify a set of views and standards for development and construction in response to the aesthetic values of architecture (Bashiri, 2011). It also seeks to create unique and interesting communities that support the values and cultures of its inhabitants. In addition, it tries to create physical environments that support the foundations of a coherent community.

Neighbourhoods with a strong sense of community actually reflect the values of their inhabitants and protect and restore unique historical, cultural, economic and geographical grounds. These neighbourhoods use natural and manmade spaces and special urban symbols in order to create a sense of belonging to the neighbourhood, communities and urban areas. Such neighbourhoods encourage constructing and maintaining buildings that are part of a local community's capital because of the services they provide and the unique contribution that they can have in creating a sense of belonging to a neighbourhood or community. Beyond the construction of buildings, such neighbourhoods help with the recognition and readability of the area for passers-by and visitors through characteristics that are unique to the area. Communities adopting the principles of smart growth can direct investments and development into areas that already reflect a strong sense of place (GSG, 2005; Lynch, 1981).

Using the perspectives on how and where communities should be, communities are able to identify and use opportunities that new developments create in accordance with distinction and beauty standards. In communities with high-quality architectural and natural elements that reflect the interests of all residents, buildings (and consequently urban neighbourhoods) are more likely to maintain their economic vitality and values over time. By doing so, infrastructure and natural resources used in creating these areas form a unique and beautiful place for residents that they will consider the home for the next generation (Rahnama & Abbas Zadeh, 2006).

The sixth principle: preserving open space, farmland, natural beauty and critical environmental areas

In recent decades, typical dispersed development patterns in the suburbs have engulfed valuable open spaces and good-quality agricultural lands. The restoration of green areas that have undergone urban development is almost impossible. Also, as the pressures from urban construction increase, protection of natural areas becomes costlier and more difficult (Seifuddini & Shorjah, 2014).

Protecting lands in suburban areas and areas with sensitive ecology can strengthen the economy of localities, empower communities, prevent degradation of the environment and culture, make outdoor leisure possible, limit sprawl and thus improve quality of life. It can also prevent severe floods and water wastage and contamination, and at the same time take care of flora and fauna (HosseinZadeh Delir & Safari, 2012).

Today, the preservation of open space is a key component in achieving better places to live.[5] Open places support the goals of smart growth by strengthening the local economic infrastructure, preserving environmentally valuable areas, providing recreational opportunities and guiding the new growth path in existing communities. Preserving open space can have a profound effect on the quality of life of a community and therefore on the region's economic boom. Open spaces and waterways can shape and guide urban forms and prevent incidental protection (reactive and small-scale protection). These "green infrastructures" can place the growth in efficient places in terms of cost[6] (Seifuddini & Shorjah, 2014).

Overall, the protection of open spaces has many economic, environmental, cultural and health benefits (HosseinZadeh Delir & Safari, 2012). The protection of valuable open spaces and environments leads to some significant health benefits in urban neighbourhoods. It has many economic benefits, including increased local assets (and hence increased property rent price) and increased revenues from tourism and local tax cuts (as a result of a decreased need for new infrastructure). Proper management and maintenance of open spaces guarantee the availability of cheap agricultural lands and valuable pastures, prevent flood damage and provide a natural and cheaper way to provide clean drinking water (Rahnama & Abbas Zadeh, 2006). In addition, the availability of open spaces increases environmental quality and health benefits. Open spaces help preserve habitats for animals and plants and maintain beautiful natural places and farmlands by eliminating development pressure and redirecting new developments into existing communities. Another benefit is the protection of the environment by dealing with air and noise pollution, controlling the wind and erosion and adjusting the temperature. In addition, open spaces support surface and underground water resources through drainage and treatment of wastewater and chemical pollutants before they get into the water system (Seifuddini & Shorjah, 2014). On the other hand, proper planning and maintenance of parks in communities pursuing smart growth is necessary (Smart Growth Network, 2002).

Seventh principle: strengthening and directing development towards existing communities

In the aftermath of World War II, urban neighbourhoods that had experienced rapid expansion in the peripheries often reflected a reduction in capital in the central parts of cities. The formation of the first suburban rings with newer development properties, such as lower density and more fragmented city fabric, led to a disregard of the inner-city context. This development model had a significant impact on the social and economic sustainability in the central parts of cities (Simons, 1998). It also led to significant effects on the environment, the occupation of open land, plant and animal habitats, water resources and their quality, air quality and global climate change (Blaha, 2000).

Thus, one of the most important goals of smart growth – formed in order to prevent the destruction and depletion of the urban centres – is to use existing infrastructure and facilities in cities to create development. This will preserve open spaces and result in a shorter work–home distance, increased efficiency of existing facilities and land, higher use of potentials and reduced development in suburban areas.

The process of increasing development in existing communities can maximize the use of impervious surfaces and thereby increase local and regional water quality. It can also create more transit options and reduce trips. Most of the existing neighbourhoods can use internal and infill development towards the re-development of brown lands and renovation of existing buildings in their agenda (Smart Growth Network, 2002, 2006).

However, re-development in existing communities faces obstacles such as detailed zoning plans, government regulations and tax subsidy policies that encourage development in marginal areas and green spaces (Cervero & Hansen, 2000). In fact, the ease of development in fertile lands around cities is a powerful deterrent to strengthening further development in existing urban areas. Marginal developments are more attractive for developers because of ease of access, lower costs of land and the possibility of construction in larger land lots (Smart Growth Network, 2002). Current zoning requirements are easier to meet in suburban areas and face fewer limitations from the existing buildings and protesting residents. Finally, the cost of land development in the periphery of cities is reduced by the government through the provision of road networks, wastewater drainage and utilities (USEPA, 2011).

In recent years, efforts have been made to reverse this tendency by increasing the motivation and passion for existing neighbourhoods; increasing people's awareness of the economic, social and environmental benefits of development in inner cities; allocation of legal incentives; etc. to steer funds towards repair and re-use of old contexts in cities (HosseinZadeh Delir & Safari, 2012).

The eighth principle: providing a variety of transportation choices

One of the problems in today's cities that has strongly affected residents' quality of life is the increase in traffic, which in turn leads to increased pollution and travel time as a result of the separation of urban uses. In addition, despite the increase in the capacity of city streets, people tend to make greater use of vehicles in order to access newly developed infrastructure (Burchell, Downs, McCann, & Mukherji, 2005). In the current situation, many local communities are designed so that their inhabitants are almost entirely dependent on personal vehicles. Residents with no transportation choices are forced to use their own cars for every trip within the city for going shopping and going to school, parks, etc.; however, new roads face traffic soon after they are developed.[7] Thus, as a result of new large roads, people will use cars more to benefit from the new infrastructure.

Studies indicate that 60 to 90 percent of the capacity of new roads – especially main roads – is filled during the first five years after their opening. In the short term, people are encouraged to use new roads instead of using public transit or car-pools, and in the long run, with the increase in access to adjacent lands, the pattern of development is directed towards expanding more and more traffic in the area. In such conditions, continuing current policies and practices cannot help to lighten the urban traffic (Smart Growth Network, 2002).

Transportation decisions affect economic, social and urban conditions of cities, both directly – by determining which land will be used for facilities and transit services such as roads, parking and entrances to cities – and indirectly – through its impact on the relative availability and development costs (Zhang, Xu, & Li, 2009). Generally, policies that reduce the costs of travelling by car (such as financial costs, travel time and difficulty) result in increased traffic and sprawl in the urban spatial structure. Table 1.8 represents those policies encouraging sprawl and those support-ing smart growth.

Often, the aim of planning decisions is to create a balance between movement (physical movement of people and goods) and accessibility (ability to achieve the desired goods and activities). This approach supports traveling by car and reduces the impact and benefits of other modes of transportation. That is because the number of large parking lots and wide streets create landscapes that reduce the possibility of walking and biking. The increase in the amount of land needed for development and the need for roads and parking lots lead to development in the outskirts, where

TABLE 1.8 Transportation policies for supporting sprawl vs. smart growth

Policies encouraging sprawl	Policies encouraging smart growth
– Increasing the capacity of streets and vehicle traffic speed	– Reducing the capacity of streets and mounting traffic speed
– Increasing the need for large-scale parking	– Reducing parking lots and their optimal management
– Reducing costs related to motor vehicles	– Allocating tolls for motor vehicles according to distance
– Reducing public transit services	– Improving transportation services and bonus strategies
– Shortage of walking and cycling paths	– Improving bike and pedestrian paths
	– Reducing traffic speed
	– Increasing network availability and management

Source: Seifuddini & Shorjah, 2014

land prices are lower. As a result, dependence on cars can create a utility for some social groups. This approach makes movements easier, but limits the alternatives and increases sprawl and car dependency. In fact, transportation by car requires relatively large amounts of land for roads and parking purposes, which reduces the amount of land per capita needed for other activities and leads to more urban sprawl.

In fact, one of the key goals of smart growth is to create a variety of transportation options so that people can use a variety of transportation systems (public transit, personal car, bikes), according to their location and conditions. This requires the precise and proper design of connections between transportation networks and facilities provided therein, as well as between land use and transit systems.

In the smart growth approach, neighbourhoods use new ways of transportation planning, such as better coordination between land use and transit systems; increased access to quality transit; different transportation options; flexibility and connections within transit networks; and connections among pedestrians, bikers and public transit and facilitate (Smart Growth Network, 2002). Based on global experiences and the results of previous studies, it seems that the only way we can reduce dependence on personal vehicles is to design convenient and efficient transportation alternatives. In addition, many people cannot drive or do not have personal vehicles. Such transit options create local communities where the elderly, youth under the age of driving and people with disabilities can easily afford to live (Bashiri, 2011).

One of the most important strategies for reducing the use of personal cars is to enhance the physical density of urban spaces, because decreasing distances provides better access to many destinations without the use of personal cars (Smart Growth Network, 2002).

Sheehan also points out that there is a clear relationship between transit and land use policies. According to him, the construction of highways in America increased the potential for the development of residential and commercial uses at the intersection of roads, and later became a stimulus for using personal cars. Thus, developers thought of roads as a great blessing. The highways provided a good potential for all arrivals and departures to other parts of the access network (Sheehan, 2001).

Sheehan (2001) believes that for better public access to transportation options, public managers can revise zoning laws and regulations to allow a mixture of residential and commercial uses and direct new developments towards places with easier accessibility to public transport systems and safer and more attractive streets for pedestrians and cyclists. However, they must also ensure easy network connections between bikes, subways, buses and other forms of transit parallel to each other (Sheehan, 2001, p. 42).

In fact, when sustainable human-oriented transportation policies are properly implemented and enforced, streets will be modified and restored, providing a safe walking route to school for children −one of the most vulnerable social groups (Sheehan, 2001, p. 42). However, this principle can only be properly realized if other principles of smart growth (such as mixed uses) are implemented simultaneously in order to maximize the effectiveness of this approach as a whole (Theart, 2007, p. 16).

In sum, it seems that replacement of car-driven development with transit-driven development requires careful pre-planning which sets transportation standards and determines the rate of tolls for roads as well as taxes for municipal parking lots. This type of planning, which supports walking and cycling, should focus on creating continuous pedestrian and bicycle routes on an urban scale, providing adequate space for parking bicycles, strengthening the quality and attractiveness of paths, increasing the compactness of urban contexts, mixing uses, increasing levels of accessibility of urban public spaces, etc. It should also consider benefits such as savings in fuel consumption and its costs, reduced emissions and finally a favourable impact on the environment and the health and vitality of residents.

The ninth principle: making development decisions predictable, fair and cost-effective

The relationship between the private and public sectors is considered crucial in smart growth development. Successful implementation of smart growth requires the effective engagement of the private sector. In fact, active participation and investment by the private sector is very important and decisive in determining the success or failure of any new development strategy (HosseinZadeh Delir & Safari, 2012). If investors, bankers, contractors, developers and the like do not see a profit, few smart growth projects will be implemented (Seifuddini & Shorjah, 2014). Expansion of smart growth requires local governments to follow development rules that are updated, justified in terms of cost and functionality and have better predictability for developers. By creating the right environment for innovative projects, orientation towards increasing foot traffic and mixed uses, the state can manage smart growth in such a way that they can benefit from more support on the part of the private sector (Smart Growth Network, 2002). To this end, developers, bankers, constructors and private-sector investors must be convinced that the approval of the proposed development changes helps them achieve their financial interests. Local governments can contribute to such confidence via the provision of financial and regulatory incentives.

Moreover, creating higher certainty and accelerating the approval process for smart growth projects is very important for builders, because the rate of return on investment is very important for investors. In other words, because builders pay high interest rates for loans, the delay in the construction approval process increases development costs. Thus, improving public–private sector coordination to support smart growth projects can help with better implementation of such projects (Seifuddini & Shorjah, 2014).

Finally, local governments must provide full support and monitoring for smart development plans to ensure their success and psychological and financial security for related private-sector investments. This can guarantee the realization of alternatives for developers in the project.

The tenth principle: encouraging community and stakeholder collaboration in development decisions

Cooperation with the public in communities where they live and work can help better explain and identify their needs. If the citizens do not participate in development projects, at best, limited investments are made and smart growth is not fully realized. In the worst cases, communities emerge that are deficient in terms of various social, economic and cultural aspects.

Development measures can create great places for living, working and leisure, as long as they can respond to the feelings of a community about how to grow and what goals to achieve. Such a response is associated with many challenges because these development measures should reflect a wide range of development needs, including the wants of community members (Twaddell & Emerine, 2007).

In fact, achieving a good prospect for a community requires the integration of the diverse needs of its members as shareholders of the community. This will include a wide range of stakeholders, including groups such as developers, urban planners and designers, transportation engineers, environmental activists, social activists, non-governmental organizations (NGOs), researchers and academics, city officials, organizations active in the field and urban management, etc. Based on this principle of smart growth, the prospects for each of these groups should be considered in the development system and proposals.

Such approaches are essential for creating mixed-use, compact, walkable and transit-oriented neighbourhoods that support smart growth. The means used in the neighbourhoods and the stakeholders are quite diverse so that a range of primary and secondary stakeholders is involved in urban planning projects. A high level of public awareness can help ensure that social needs and possible solutions have been fully considered. This strategy can help local managers recognize and support the kind of development that best matches the actual needs and priorities of the local community (USEPA, 2011).

In this regard, smart growth can create desirable places to live, work and have fun. However, as communities have different needs, finding maximum reasonable benefits for the parties involved and the beneficiaries is sometimes very difficult. Those with a stronger economic status may expect faster methods of construction and housing development, whereas those individuals and groups lacking capital may prefer infill developments. New communities with distant and detached uses may seek a sense of place through the development of mixed uses in areas of urban centres, whereas those people who are suffering with poor air quality demand measures to solve this problem. A conclusion that can be drawn from these demands is that the needs of each community and the programs it follows can best be defined by the people who live and work in it. Public participation can – in many cases – be time consuming, frustrating and expensive. However, strengthening cooperation among different stakeholders can lead to faster and more creative achievement of development results and show the importance of public awareness about proper investment and planning. Smart growth development policies and programs without strong investment and participation by citizens will not be robust, at best, and, at worst, they will lead to poor communities. When people feel they are unaware of the importance of rules, they will not try to comply with them. Higher public participation in the planning process improves public support for smart growth and often leads to new strategies that are relevant to the specific needs of the community (Smart Growth Network, 2002).

The principles of smart growth ensure that any new approach in development can actively and equitably satisfy the needs of all stakeholders. The basis for smart growth should be identified and supported by recognizing the specific short- and long-term goals of the community based on measures of community quality of life.

The discussion of the ten principles of smart growth covered here is context free, and they can be prioritized and implemented in every community, depending on its circumstances and needs. Table 1.9 shows these ten principles in relation to the main urban components (such as the natural environment, community, economy and urban management).

TABLE 1.9 The effects and results of applying smart growth principles in relation to the main urban components

Urban components	Smart growth principles	Results
Natural environment	Use of compact development model	– More efficient use of land and resources. – The possibility of creating more open spaces with a decrease in new construction. – Protecting undeveloped open spaces, promoting surface runoff treatment, reducing the need for drainage and reducing the pollution of rivers and lakes. – Reducing development pressures on open spaces and agricultural lands. – The possibility of creating fields, playgrounds, gardens and lands for recreation and exercise in the remaining open spaces. – Reducing heat islands as a result of a reduction in the amount of land dedicated to pathways and associated facilities. – Supporting various transit options within the city and cutting the related costs. – Placing essential services within walking distance and reducing the reliance on cars and the demand for parking spaces. – Increasing the economic value of buildings relative to urban developments in the suburbs.

Urban components	Smart growth principles	Results
	Strengthening and directing the development towards existing built contexts (use of lands within the city)	– Increasing the productivity of existing lands and infrastructure. – Protecting open space and non-renewable natural resources around cities. – Enhancing the quality of life and affordable growth. – Benefiting from a wider range of jobs and services in close proximity to each other. – Maintaining more open spaces and in some cases strengthening rural communities. – Reducing development pressures away from the open spaces and good-quality agricultural land in the peripheries. – Proximity of residential and working areas, thus shortening commuting trips.
	Protection of open space, farmland, natural beauty and critical environmental areas	– Helping with surface runoff adsorption and treatment. – Reducing the pollution of rivers and lakes. – Preventing damage to urban areas due to floods and surface runoff by reducing impervious surfaces. – Reducing the need for drainage. – Improving air quality. – Improving the quality of life. – Controlling and guiding development in a planned way.
Community	Creating a range of housing opportunities and choices	– Supporting different demographic groups by integration of single- and multi-family structures. – Fair distribution of households with different income levels. – Creating economic impetus for commercial centres by constructing new housing.
	Creating walkable neighbourhoods	– Increasing social interaction in urban areas. – Increasing the sense of security in neighbourhoods. – Reducing dependence on cars and, in turn, improving air quality. – Reducing the need for large parking lots. – Providing services to a wider range of users (including pedestrians, cyclists and others).

(*Continued*)

TABLE 1.9 (Continued)

Urban components	Smart growth principles	Results
	Providing a variety of transportation choices (offering various movement options such as walking, cycling, public transit and private cars)	− Reducing the amount of land used for transportation − Reducing citizens' reliance on cars. − Avoiding traffic congestion. − Improving air quality. − Increasing the accessibility of high-quality transit services to the public. − Equitable distribution and optimal use of land in accordance with the public transit network. − Optimal spatial distribution of key urban uses in order to increase the likelihood of navigating shorter distances on foot that can promote citizens' health. − Increasing the vitality of urban spaces. − Helping create population centres and urban centres active during different hours of the day.
	Fostering distinctive, attractive communities with a strong sense of place (valuing the existing neighbourhoods)	− Supporting the values and cultures of the community residents. − Using those natural and architectural elements that reflect the interests of residents. − Maintaining the vitality and economic value of the neighbourhoods. − Improving the quality of life.
	Mixing land uses	− Increasing the use of public transit instead of cars. − Attracting a diverse population. − Enhanced neighbourhood security as a result of the increased presence of people on the streets. − Increased inter-personal interaction and, in turn, social revitalization. − Higher support for the development of essential services using the existing infrastructure. − Economic stability.
Urban economy and management	Making predictable, fair and cost-effective development decisions	− Providing the high costs required for development based on smart growth by large private markets. − Making smart growth profitable for the private- and public-sector developers by making the correct decisions and implementing the right regulations.

Urban components	Smart growth principles	Results
	Encouraging community and stakeholder collaboration in development decisions	− Satisfying residents by considering their development needs. − Creating a better community to live and work in. − Rapid solutions to development issues and a better understanding of the importance of good planning. − Promoting public support for smart growth. − Achieving innovative strategies.

Source: Adapted from: (Smart Growth Network, 2002; Smart Growth Network, 2006; Alexander & Tomalty, 2002; Theart, 2007)

Additional principle: the need for collaboration among different players

Besides the ten principles introduced by the SGN, some scholars argued that smart growth should bring different players (such as government, city authorities, public and private sectors) in multi-scale areas (regional, city, local) around a table. On one side, smart growth is clearly related to the new regionalism, as some of the policies, such as tax base sharing, and planning tools, like UGBs, are only manageable on a region-wide scale (Counsell & Haughton, 2004). At the same time, smart growth pays particular attention to local community aspects by emphasizing the importance of the socio-economically balanced territorial network and of high physical quality in the local built environment (Calthorpe & Fulton, 2001). In so doing, it emphasizes the social roots of the concept of sustainability by highlighting the role of the local bodies in designing their future (Beatley & Manning, 1997; Trillo, 2013) and the commitment of local governments to achieve consistent and effective land use patterns, thus implicitly recognizing how in the age of "decentralized leadership" "all sustainability is local", as "the leaders who shape the built environment . . . number in the millions of people" (Farr, 2011, p. 10).

The impact of smart growth on urban life

Given the discussions presented in this chapter, we can say that smart growth is not fixed and inflexible and can take different approaches depending on the sample. Thus, smart growth can have a tremendous effect on cities and urban areas. In this section, the effects of smart growth − both the positive and negative aspects − are examined. The aim is to achieve a precise understanding of the impacts of smart growth on cities. These effects are sometimes in conflict with each other, but sometimes complement each other as well.

One of the strategies used in smart growth is that of *integrated transportation planning and land use* (Litman, 2015b). For example, smart growth tries to create a coherent network of sidewalks, biking paths, etc. in the vicinity of schools, malls and commercial areas. Besides the design for pedestrians or bicycle users, the public transit system is enhanced and expanded in these areas. This strategy helps citizens who do not use personal vehicles or do not have access to them. Such a strategy in the form of a good plan can increase economic efficiency in the commercial centres of a city (Cervero, 2002; Handy, 2005).

Efficient positioning of future developments is another important smart growth strategy (Litman, 2015b) that can have the tangible effect of providing affordable housing with suitable access points. Such a strategy satisfies the housing needs of low-income people and is an appropriate response to the demand in the housing and urban facilities market. It increases the economic efficiency of cities and prevents the formation of poor forms of housing such as slums or shantytowns (Litman, 2010).

Another smart growth strategy is *increasing flexibility in the functional zoning of the city* (Litman, 2015b). This strategy allows more compact urban forms and development based on mixed uses. In fact, combining different uses in urban areas makes the city more dynamic and active. The variation in the land use benefits all inhabitants and increases social interactions that, in turn, lead to a better quality of life. Research shows that though mixing uses does not have the benefits of aggregation of a group of specific uses, it does help distribute and balance the economic interests of all uses and prevents the formation of specific hubs in cities. For example, the formation of a commercial hub leads to many problems such as heavy traffic during the day, and an administrative hub becomes completely devoid of people at the end of the day and during the night. Thus, the logical combination of uses can provide convenience in terms of transit and efficiency in economic performance (Molinaro, 2003).

Urban growth control, which limits the growth boundaries of a city and prevents development in natural areas, is among the main objectives and strategies of smart growth (Litman, 2015b). Such a strategy has always faced strong opposition (Litman, 2015a). Citizens who prefer to live in an open space in single-family housing disagree with such an approach and deem urban growth control against their own property rights. Limiting city growth reduces dependence on personal vehicles that, in turn, brings about many desirable social, economic and environmental outcomes.

Finally, *change in transportation budget* is another smart growth strategy that aims to reduce the budgets used for the construction of highways and freeways in favour of funding infrastructure for pedestrians, bicycles and public transit (Litman, 2015b). People using transportation alternatives instead of personal vehicles form the most important target group of this strategy. Although the capacity for personal vehicles decreases or, at least does not continue to develop, proper planning for various modes of public transit reduces vehicle traffic and satisfies the needs of users of private vehicles as well (U.S. Environmental Protection Agency, 2012a).

This strategy increases the economic efficiency of public transport; reduces pollution due to improper methods of transportation; and reduces traffic, accidents, and fatalities.

From another perspective, the impact of smart growth can be classified and investigated in three categories: urban economy, urban communities and urban environment. In the following sections, each of these cases is discussed in turn.

Urban economy

In recent years and in advanced societies, people are looking for dynamic and diverse places to live and work. These people want different options for housing and transit and tend to have daily trips on foot or by bike for their personal needs (Litman, 2016). Entrepreneurs want to locate their activities in areas that can attract more people and are available for their employees (U.S. Environmental Protection Agency, 2012b). Local authorities are eager to attract people and capital to their urban areas and try to limit the resources used for the development of urban infrastructure (U.S. Environmental Protection Agency, 2012b). The private sector and mass constructors try to increase their profit by stimulating market demand (Hess, 2016). Smart growth provides opportunities to meet all these needs by linking economic development with public property and infrastructure, thereby creating attractive places for the business sector and citizens alike.

The principles of smart growth can create a space with a stronger local economy and help protect the environment. A study in 2004 entitled "Smart Growth Is Smart Business" showed that return on investment in existing communities is higher, as it faces lower costs compared to new developments and returns more short-term and long-term economic benefits to the investor (Smart Growth Leadership Institute, 2004). Smart growth does not hinder economic competitiveness in the market; therefore, the free market can meet consumer demand in the inner regions of the city and gain benefits from the smart growth market (Simmons, 1999). Also, smart growth with the compact city plan at its heart can reduce the costs of the public and private sectors, and the resulting savings can be used in other areas of the city, such as in the creation of leisure spaces and public transit. This can enhance the quality of life and attract capital and people to the city. Finally, smart growth reduces the costs of housing and transit that has high economic utility for many citizens, who will pay lower prices for housing, travel shorter distances and will have more leisure time (Shapiro, 2005). The economic domain of smart growth starts with citizens and satisfies investors and the private sector, thereby reducing the financial burden on the public sector and government (U.S. Environmental Protection Agency, 2012a).

Urban community

The impact of smart growth on the urban community and citizens can be discussed from two perspectives: quality of life and human health.

Quality of life depends on many factors and includes a variety of indicators and metrics. One of these indicators is *the citizens' access to housing* (Holloway & Guy, 2000; Preuss & Vemuri, 2004; Shapiro, 2005). Smart growth gives priority to affordable housing in an attempt to provide citizens with good housing with convenient access (Litman, 2010). Another important indicator of the quality of urban life is *the cost of housing and life* (Preuss & Vemuri, 2004; Shapiro, 2005). Smart growth creates compact patterns and multi-family housing to reduce the share of housing costs in the cost of living (Litman, 2015a). However, it adds to taxes by increasing urban facilities and equipment and their balanced distribution in all urban areas. Whether this increase and decrease in various parameters will be to the benefit of all citizens is not discussed in this work. However, these examples do not cover all living costs of citizens and only represent two different aspects of the impact of smart growth on this factor.

Another indicator is *the population a city attracts* (Holloway & Guy, 2000; Preuss & Vemuri, 2004). The higher rate of received population in a city as compared to its depopulation rate is considered an indicator of a high quality of life. However, research has found that the immigration status of smart cities is not a function of smart growth. Instead, the dynamics of a city that is the outcome of smart growth strategies, such as mixing uses, have diverse neighbourhoods and providing efficient transit, is important in attracting populations (Preuss & Vemuri, 2004).

Reducing the home-to-work commuting distance is another factor that affects the quality of life (Geller, 2003; Holloway & Guy, 2000; Preuss & Vemuri, 2004). In this regard, smart growth strategies of mixed use, more compact urban areas, walkability and the possibility of biking are relevant. *Reducing traffic congestion and air pollution* is another indicator of quality of life to which smart growth can respond by shifting from car-oriented to human-oriented cities. Although quality-of-life measures are more than what was mentioned earlier, based on the available studies, it can be concluded that smart growth can have a significant impact on quality of life (U.S. Environmental Protection Agency, 2012c).

Another important impact of smart growth on the urban community relates to *human health* (Dannenberg, Jackson, & Frumkin, 2003; Frank, Kavage, & Litman, 2006). The traditional method of planning communities affects public health directly. Conventional housing designs with wide streets and no sidewalks increase the use of cars even for short trips. Many settlements are designed in a way that increases the use of cars, and their pedestrian crossings and passages are used for pedestrians and bicycle traffic (Dalbey, 2008). These patterns of development lead to sedentary behaviours through increased reliance on cars. Research shows that these growth patterns increase obesity among children and adults. Today, one in four Americans is obese and 60 percent of Americans suffer from overweight. Also, during the past three decades, the percentage of overweight children has doubled (Eida, Overman, Puga, & Turner, 2008; Ewing, Schmid, Killingsworth, Zlot, & Raudenbush, 2004).

Smart growth is a perfect solution that can increase public health by identifying and controlling sources of environmental pollutants, improving the safety of pedestrians, forcing residents and staff towards higher levels of activity and creating a healthier and higher quality of life. So we can say that smart growth has significant impacts on human communities that are – in most cases – in favour of the urban population (Dannenberg et al., 2003; Holloway & Guy, 2000).

Urban environment

The decisions made to locate housing, offices, streets and other pillars of development have long-term effects on the environment (Talen & Knaap, 2003). Smart growth attempts to achieve cleaner air, higher-quality water and better protections for nature. Smart growth seeks the conservation of nature alongside better protection of social and economic life. For example, the location of shopping malls, schools and businesses near residential areas reduces the length of travel between these components. The shorter the distance travelled, the lower the air pollution in neighbourhoods and urban centres, and the lower amounts of greenhouse gases released into the urban space (Filion & McSpurren, 2007; Miller & Hoel, 2002).

Buses, trains and subways reduce the share of private car traffic, as well as accidents and air and noise pollution. Emissions from vehicles, including carbon monoxide, nitrogen dioxide, ozone and fine particulate matter, impose heavy costs on human health, as well as flora and fauna. Recovering from such losses would cost billions for governments. Although new technology promises to produce cars with lower emissions in the coming years, the simplest and least expensive solution for now is to reduce reliance on personal cars (Filion & McSpurren, 2007).

On the other hand, the amount of artificial and built surfaces is inversely related to the amount of water absorbed by the ground (Johnson, 2001). In fact, such surfaces create an insulating layer against the permeation of water; therefore, runoff is either forced out of the natural cycle through evaporation or forms floods. That is why many cities face water shortages despite high precipitation rates.

The increasing development towards the outskirts devours natural areas and agricultural lands and destroys many valuable natural landscapes such as forests and wetlands. It disrupts the performance of these natural spaces, leading to lack of flood control, migration of animal species and the disappearance of certain species of plants. The growth of cities and built areas robs the cities of farmlands and leads them towards the use of low-quality soils. In such conditions, the destruction of forests and wetlands increases and dependence on irrigation and fertilizers is intensified (Hwang & Foster, 2006; Johnson, 2001; Wu, 2006).

In the previous section, the wider economic impacts of smart growth in connection with citizens and the private and public sectors were discussed. Also, important issues in terms of quality of life and citizens' health were discussed for the urban community, and the effects of smart growth on these two factors were studied. Finally, the effects of smart growth on the most important concern in the third millennium (i.e. the environment) were investigated. The following section summarizes the costs and benefits of smart growth.

A brief look at the benefits of smart growth

This section summarizes the advantages and disadvantages of smart growth. A useful factor in evaluating the benefits includes the costs and consequences of urban sprawl. According to research on the damages of urban sprawl, the costs can be divided into two general categories (Burchell, 2000; Burchell et al., 1998; Carruthers & Ulfarsson, 2002; Gordon & Richardson, 2000; Litman, 2015a, 2015b). The first is the per capita consumption of land. As is obvious, higher consumption of land increases the costs of providing public services and infrastructure. As a result, farmlands are lost, food production decreases and natural resources are damaged. Measures to offset or reduce these adverse effects add to the costs of urban sprawl. The second category of costs is related to scattered activities that reduce access across the city and make the provision of a public transit infrastructure difficult and costly. Scattered activities reduce the possibility of creating pedestrian- and bike-oriented transportation systems. Also, more reliance on personal cars leads to problems such as the intense need for parking, crowds and accidents that impose heavy costs on the public sector and taxpayers. Figure 1.6 shows the effects of urban sprawl on these categories (Litman, 2015a).

Smart growth has lots of benefits compared to the costs of urban sprawl. These include affordable housing, protection of open and natural spaces, reduced costs of public services and infrastructure, multiple options and modes of transport and citizens' health and wellbeing. Appropriate implementation of smart growth strategies can have diverse economic, social and environmental benefits, some of which are listed in Table 1.10.

Smart growth usually supports economic development by enhancing economic productivity and lowering costs. Some studies indicate that use of the smart growth recommendations reduces the costs of public services, including those related to water and sewage, as well as pathways and roads (Carruthers & Ulfarsson, 2002; Litman, 2004).

Urban density is one of the most controversial issues in smart growth and is closely associated with cost reduction. City life combined with high density is prerequisite to vitality, inspiration and social and collective interactions (Song, 2005; Turner, 2006). Smart growth is less dependent on personal vehicles and reduces the need for a transportation infrastructure (Handy, 2005). It should be noted that smart growth does not seek to remove cars, but decreases car trips per person compared with urban sprawl. McCann and Ewing (2003) argue that more than 20 percent of costs of households in scattered areas belong to road transport, whereas this percentage for households living in communities with efficient space planning is below 17 percent. Thus, it can be said that households in areas with better accessibility can save significant amounts on annual costs of transportation (McCann & Ewing, 2003). An important parameter of smart growth that is directly related to

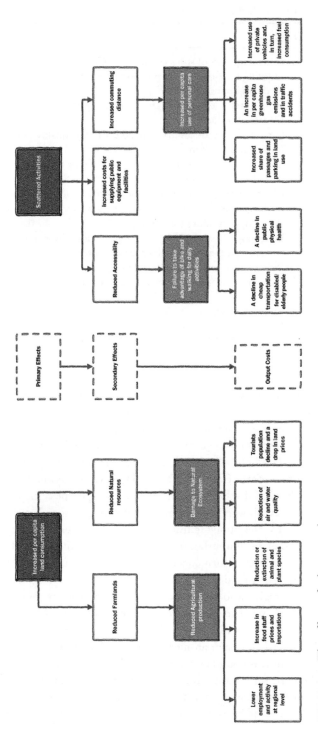

FIGURE 1.6 The effects of urban sprawl

Source: Litman, 2016

TABLE 1.10 A summary of the benefits of smart growth

Benefits	Description
Economic	– Reducing costs of services and development
	– Limiting urban development and protecting farmlands and orchards
	– Reducing transit costs
	– Economies of concentration and agglomeration
	– Efficient transportation
	– Supporting industries that depend on a high-quality environment (tourism, agriculture)
Social	– Improving transit opportunities, especially for those who do not have the ability to drive
	– Concentrating local activities and improving the quality of life in neighbourhoods; higher safety and a more active environment
	– Better opportunities for housing
	– Increasing physical activity and improving health conditions
	– Preserving unique cultural resources (historical, traditional)
Environmental	– Protecting green spaces and wildlife
	– Increasing the use of public transit and reducing environmental waste
	– Overall reduction of pollutants and greenhouse gases
	– Reducing water pollution

Source: Litman, 2015a

access and its impact on urban trips is land use density and mixing (Handy, 2005). Table 1.11 shows land use effects on people's decision to travel.

However, as with any other approach, smart growth can have some costs (Table 1.12). For example, in some cases, growth control can deprive some families of large-scale, single-family houses, and planning diverse public transit options can reduce the speed or comfort of travelling by car. However, offsetting these negative effects is possible through diversity in smaller housing types, placement of schools and shops in dense neighbourhoods, improvement of walking and cycling conditions and provision of better public transit services. To the extent that these reforms attract people to denser neighbourhoods and alternative methods, the demand for large houses reduces, which in turn cuts crowds and traffic sought by citizens who prefer scatted areas and travelling by car. Smart growth and integration of land use management strategies lead to an increase in some development costs and a decrease in other costs. In particular, costs of planning, costs per unit of land, project costs for infill construction and renovation and costs of higher design standards increase. On the other hand, such an increase in costs as a result of integration in infrastructure and transit is compensated for by reducing the amount of land tenure; reducing the need for infrastructure, roads and parking; reducing maintenance costs; and, ultimately, increasing opportunities for saving. In general (especially in the long term), the costs of smart growth will be equal to or less than urban sprawl (Litman, 2015a).

TABLE 1.11 Land use effects on people's decision to travel

Factor	Definition	Consequences for travel
Density	Number of people or jobs per unit of land measurement (hectares or square kilometres)	High density reduces vehicle trips per person
Mixing	The extent to which various land uses are placed near each other	Mixing land use reduces car trips and increases alternative transport methods.
Accessibility	Development related to urban/ regional centres	Better accessibility leads to reduced use of vehicles. For example, the urban trips by residents of inner neighbourhoods are 10 to 30 percent lower than that of suburban residents.
Concentration	The extent of business and other activities in centres	Concentration increases the use of alternative transit options.
Network connection	The degree to which the sidewalks and streets are connected to each other, leading to increased inter-city journeys.	Increased network connections reduce distances travelled by vehicles and increase walking and cycling.
Walking and biking conditions	The quantity, quality and safety of sidewalks and bicycle routes	Better conditions for cycling and walking increase intra-city trips without cars.
Preparation and management of parking	The number of spaces assigned to parking per building or per acre and the quality of their management and regulation	Reducing parking spaces, and the rate of their use reduces car trips.

Source: Strafford Regional Planning Commission, 2011

TABLE 1.12 The benefits and costs of smart growth

Internal		External
Benefits	– Improved housing options (fewer restrictions for multi-family dwellings) – Increasing the affordability of housing (e.g. reducing the need for land and parking) – Improving accessibility options – Saving transportation costs – Reducing the risk of accidents – Improving public health – More attractive and liveable neighbourhoods – Reducing the need to drive	– Saving the costs of public services (lower costs for roads, living facilities, emergency services, etc.) – Reducing costs/subsidies for roads and parking – Reduce congestion (if people drive less) – Reducing the accident risk for other road users – Increasing social cohesion – Improving accessibility for non-drivers – Conserving energy – Reducing emissions – Preserving open spaces (farms and other spaces)

(*Continued*)

TABLE 1.12 (Continued)

Internal		External
Costs	− Very small size of land plots − Less privacy − Lower speed of local traffic − Higher road and parking tolls − Payment of taxes for some local pollutants	− Additional infrastructure costs (such as expanding public transit network) − Increase in local traffic congestion − Increase in the impervious surfaces in some areas

Source: Litman, 2015a

In general, we cannot claim that all smart growth policies are useful for everyone. The degree to which smart growth can cause increased compactness and accessibility, the formation of diverse communities, reduced land consumption, less driving and higher reliance on alternative modes of transit will have a wider range of direct and indirect benefits for citizens. All these consequences should be considered when assessing the effects and consequences of smart growth. In Chapter 4 of this book the criticisms of smart growth will be discussed.

Conclusion

Williams (2003) believes that smart growth will remain impotent and like a stylish boutique in the absence of large-scale changes. However, Chen (2004) states that smart growth is still moving fast on the rails of success. According to Chen, smart growth has made remarkable achievements in terms of legislation, research, new alliances to achieve common goals and raising public awareness on a larger scale. Despite this belief, urban sprawl is still the dominant paradigm for growth worldwide, which has a negative effect on the quality and sustainability of the community. M. Harris (2004), the director of the Joint Center for Sustainable Communities, believes that despite the extensive public awareness of smart growth, its recognition and acceptance by the public are not at the level deserved. People in various countries still think of the American Dream and its certain type of residence (large-scale plots in the suburbs, commuting by private cars and so on) and have no desire to live in denser areas or near the centre of the city. However, this situation may change with the increase in oil and petrol prices. Corbett (2004) believes that smart growth cannot yet be considered a normal process and specific interventions should be made to ensure its impacts. It seems that Corbett (2004) is right. There are many policies and strategies that do not support and do not approve the principles of smart growth. Local governments and residents are not always able to recognize the benefits arising from smart growth strategies in their regions and cities. According to McElfish (2007), smart growth has a convincing power and ability arising out of the variety of choices it can present for development, improvement of communities and preservation of environmental interests in the future (Theart, 2007).

Finally, it seems that particular attention should be given to the following issues in order to advance the principles of smart growth:

- Strategies for smart growth should be further promoted.
- Principles of smart growth should form part of national, provincial and local policies and regulations.
- Governments need to have a better understanding of the opportunities that smart growth can create for them.
- More practical policies should replace long and large documents, which are very complex and difficult to implement.
- Communities should be trained to change their policies and move towards the smart growth paradigm. However, this will be achievable only if the concept is fully understood.

Notes

1 The Club of Rome was founded in 1968 by a group of 30 scientists and politicians from different countries in Rome to discuss the problems and crises around the world.
2 In fact, the theories of compact cities and smart growth are closely interrelated, because both seek more sustainable and efficient use of land to help control congestion in central areas (Cho, 2008).
3 One of the goals of smart growth is to build low-income housing for various classes in society. When different housing options are placed next to each other, households with different income levels are better distributed in the city and public services are more effectively provided to individuals. Therefore, social justice is achieved in part.
4 According to a study done in the United States, about 36 percent of urban neighborhood residents choose walking among different methods of transport, provided that there are appropriate and equipped paths and sidewalks for this purpose at their disposal.
5 Smart growth uses the term "open space" for available natural areas within cities as well as the habitats of flora and fauna, entertainment venues, farms and ranches, natural places and areas of environmental preservation.
6 The most appropriate and cost-effective location for a new development is where roads, waterways and other utilities are present. Green infrastructure also means ensuring the protection of wildlife, water quality and land efficiency in terms of sustainability and cost-effectiveness (Berkeley, 2000).
7 This is called induced demand (saturation) in the literature.

References

Abley, S., & Turner, S. (2011). *Predicting walkability* Retrieved from New Zealand: https://www.nzta.govt.nz/resources/research/reports/452/docs/452.pdf

Alexander, D., & Tomalty, R. (2002). Smart growth and sustainable development: Challenges, solutions and policy directions. *Local Environment, 7*(4), 397–409.

American Planning Association. (1997). *Growing smart legislative guidebook: Model statues for planning and management of change*. Chicago: American Planning Association.

American Planning Association. (2002). *Planning for smart growth 2002 state of states*. Chicago: American Planning Association.

American Planning Association. (2006). *Planning and urban design standards*. New York & London: John Wiley & Sons.

Anacker, K. B. (2015). *The new American suburb: Poverty, race and the economic crisis*. New York: Routledge.

Andrusz, G., Harloe, M., & Szelényi, I. (2011). *Cities after socialism: Urban and regional change and conflict in post-socialist societies*. Oxford: Blackwell.

Azizi, M. (2004). *Density in urbanization; principles and criteria for determining urban density*. Tehran: Tehran University.

Banister, D., Watson, S., & Wood, C. (1997). Sustainable cities: Transport, energy, and urban form. *Environment and Planning B: Planning and Design, 24*(1).

Bartelmus, P. (1999). Sustainable development- paradigm or paranoia? *Wuppertal Papers*.

Bartley, R. (1996). Public-private partnerships and performance in services provision. *Urban Studies, 33*(4/5).

Bashiri, L. (2011). *Determining urban residential density based on smart growth approach*. (Master of arts dissertation), Tehran Art University.

Beatley, T., & Manning, K. (1997). *The ecology of place: Planning for environment, economy, and community*. Washington, DC: Island Press.

Belzer, D., & Autler, G. (2002). *Transit oriented development: Moving from rhetoric to reality*. Washington, DC: Brookings Institution Center on Urban and Metropolitan Policy.

Berke, P., Godschalk, D., Kaiser, E., & Rodrigerz, D. (2006). *Urban land use planning*. Chicago: University of Illinois Press.

Berkeley, C. (2000). Economic & Planning Systems, Regional Economic Analysis (Trends, Year 2000 & Beyond), San Leandro: East Bay Regional Park District. Retrieved from www.ebparks.org/resources/resources.html

Bertolini, L. (1996). Nodes and places: Complexities of railway station redevelopment. *European Planning Studies, 4*(3), 331–345.

Bertolini, L., & Spit, T. (2005). *Cities on rails: The redevelopment of railway stations and their surroundings*. London: Routledge.

Blaha, K. H. P. (2000). *Opportunities for smarter growth: Parks, greenspace and land conservation translation*. Miami: Collins Center for Public Policy.

Bohl, C. (2000). New urbanism and the city: Potential applications and implications for distressed inner-city neighborhoods. *Housing Policy Debate, 11*(4).

Bohl, C. C., & Schwanke, D. (2002). *Place making: Developing town centers, main streets, and urban villages*. Retrieved from Washington, DC.

Borba, A. (2014). Cycling Amsterdam. In Cycling_Amsterdan_04 (Ed.): Wikimedia.Org.

Bourne, L. S. (1992). Self-fulfilling prophesies?: Decentralization, inner city decline, and the quality of urban life. *Journal of the American Planning Association, 58*(4), 509–513.

Breheny, M. (1995). The compact city and transport energy consumption. *Transactions of the Institute of British Geographers*, 81–101.

Bunce, S. (2004). The emergence of 'smart growth' intensification in Toronto: Environment and economy in the new official plan. *Local Environment, 59*.

Burchell, R. (2000). *Costs of sprawl 2000*. Washington, DC: National Academy Press.

Burchell, R., Downs, A., McCann, B., & Mukherji, S. (2005). *Sprawl costs: Economic impacts of unchecked development*: Island Press.

Burchell, R. W., Shad, N. A., Listokin, D., Phillips, H., & Downs, A. (1998). *The costs of sprawl – revisited*. New Brunswick: Rutgers University.

Burton, E., Jenks, M., & Williams, K. (2003). *The compact city: A sustainable urban form?* London: Routledge.

California Department of Transportation. (2005). *Transit-oriented development compendium*. Retrieved from California.

Calthorpe, P. (1989). *The pedestrian pocket book*. New York: Princeton Architectural Press.

Calthorpe, P. (1993). *The next American metropolis: Ecology, community, and the American dream.* New York: Princeton architectural press.

Calthorpe, P. (1994). The next American metropolis. *Architectural Design,* (108), 18–23.

Calthorpe, P., Corbett, M., Duany, A., Moule, E., Plater-Zyberk, E., & Polyzoides, S. (1991). *The Ahwahnee principles.* Retrieved from www.lgc.org/wordpress/docs/ahwahnee/ahwahnee_principles.pdf

Calthorpe, P., & Fulton, W. (2001). *The regional city.* Washington, DC: Island Press.

Campbell-Lendrum, D., & Corvalán, C. (2007). Climate change and developing-country cities: Implications for environmental health and equity. *Journal of Urban Health, 84*(1), 109–117.

Carruthers, J. I., & Ulfarsson, G. F. (2002). Urban sprawl and the cost of public services. *Environment and Planning B, 30*(4), 503–522.

Cervero, R. (1996). Travel choices in pedestrian versus automobile oriented neighborhoods. *Transport Policy, 3*(3).

Cervero, R. (2002). Induced travel demand: Research design, empirical evidence, and normative policies. *Journal of Planning Literature, 17.*

Cervero, R. (2004). *Transit-oriented development in the United States: Experiences, challenges, and prospects* (Vol. 102). Washington, DC: Transportation Research Board.

Cervero, R., Ferrell, C., & Murphy, S. (2002). Transit-oriented development and joint development in the United States: A literature review. *TCRP Research Results Digest* (52).

Cervero, R., & Hansen, M. (2000). *Road supply – demand relationships: Sorting out casual linkages.* California: University of California Transportation Center.

Cervero, R., & Kockelman, K. (1997). Travel demand and the 3Ds: Density, diversity, and design. *Transportation Research Part D: Transport and Environment, 2*(3), 199–219.

Chaolin, G. (1994). *West schools of urban study after the second world war.* Retrieved from Beijing.

Chen, D. (2004). What is the state of smart growth today, getting smart. *Newsletter of the Smart Growth Network.*

Cho, K. (2008). *Fiscal impact analysis for a smart growth zoning strategy: A study of west campus university neighborhood overlay district.* (Master of Science), The University of Texas, Austin.

Cleaver, H. M. (1982). The contradictions of the green revolution. *The American Economic Review, 62.*

Cohen, Steve (2018). *California's misguided attempt to force urban density.* Retrieved from New York.

Corbett, J. (2004). What is the state of smart growth today, getting smart. *Newsletter of the Smart Growth Network.*

Counsell, D., & Haughton, G. (2004). *Regions, spatial strategies and sustainable development.* London: Routledge.

Currie, G. (2006). Bus transit oriented development – strengths and challenges relative to rail. *Journal of Public Transportation, 9*(4), 1.

Dalbey, M. (2008). Implementing smart growth strategies in rural America: Development patterns that support public health goals. *Journal of Public Health Management & Practice, 14*(3).

Dannenberg, A. L., Jackson, R. J., & Frumkin, H. (2003). The impact of community design and land-use choices on public health: A scientific research agenda. *American Journal of Public Health, 93*(9).

Dantzig, G. B., & Saaty, T. L. (1973). *Compact city: A plan for a liveable urban environment.* San Francisco: W. H. Freeman.

Davey, K. (1996). *Urban management: The challenge of growth.* Avebury: Aldershot.

Davidson, F., & Payne, G. (2000). *Urban projects manual.* Liverpool: Liverpool University Press.

Department of Infrastructure, P. a. E. (2012). *Roads and the environment*. Retrieved from Australia: https://transport.nt.gov.au/publications

Dittmar, H., & Ohland, G. (2012). *The new transit town: Best practices in transit-oriented development*. Washington, DC: Island Press.

Dong, H. (2010). *Assessing Portland's smart growth: A comprehensive housing supply and location choice modeling approach*. Portland: Portland State University.

Downs, A. (2005). Smart growth: Why we discuss it more than we do it. *Journal of the American Planning Association, 74*(4), 367–380.

Drakakis-smith, D. (1995). Third world cities: Sustainable urban development. *Urban Studies, 32.*

Eida, J., Overman, H. G., Puga, D., & Turner, M. A. (2008). Fat city: Questioning the relationship between urban sprawl and obesity. *Journal of Urban Economics, 63*(2).

Elkin, T., McLaren, D., & Hillman, M. (1991). *Reviving the city: Towards sustainable urban development*. London: Friends of the Earth Trust.

Environmental Protection Agency. (2013). *Strategies for advancing smart growth, environmental justice, and equitable development*. Retrieved from USA.

Ewing, R. (1999). *Pedestrian and transit friendly design: A primer for smart growth*. Washington, DC: International City/County Management Association and Smart Growth Network.

Ewing, R., Hamidi, S., & Grace, J. B. (2016a). Urban sprawl as a risk factor in motor vehicle crashes. *Urban Studies, 53*(2).

Ewing, R., Hamidi, S., Grace, J. B., & Weid, Y. D. (2016b). Does urban sprawl hold down upward mobility? *Landscape and Urban Planning, 148.*

Ewing, R., Schmid, T., Killingsworth, R., Zlot, A., & Raudenbush, S. (2004). Relationship between urban sprawl and physical activity, obesity, and morbidity. *Urban Ecology, 18*(1), 47–57.

Farr, D. (2011). *Sustainable urbanism: Urban design with nature*. New York & London: John Wiley & Sons.

Ferrier, K., Caves, R., & Calavita, N. (2016). The challenges of smart growth: The San Diego case. In F. W. Wagner, T. E. Joder, A. J. Mumphrey, Jr., K. M. Akundi, & A. F. J. Artibise (Eds.), *Revitalizing the city: Strategies to contain sprawl and revive the core* (pp. 59–86). London: Routledge.

Filion, P., & McSpurren, K. (2007). Smart growth and development reality: The difficult co-ordination of land use and transport objectives. *Urban Studies, 44*(3).

Fleissig, W., & Jacobsen, V. (2002). *Smart scorecard for development projects*. Paper presented at the Congress for New Urbanism and US Environmental Protection Agency.

Frank, L., Kavage, S., & Litman, T. (2006). *Promoting public health through smart growth: Building healthier communities through transportation and land use policies and practices*. Retrieved from United States of America.

Frece, J. W. (2008). *Sprawl & politics the inside story of smart growth in Maryland*. Albany: State University of New York Press.

Galster, G., Hanson, R., Ratcliffe, M. R., & Woldman, H. (2001). Wrestling sprawl to the ground: Defining and measuring an elusive concept. *Housing Policy Debate, 12*(4).

Geller, A. L. (2003). Smart growth: A prescription for livable cities. *American Journal of Public Health, 93*(9).

Gillham, O. (2002). *The limitless city: A primer on the urban sprawl debate*. London: Island Press.

Girardet, H. (2004). The metabolism of cities. In S. M. Wheeler & T. Beatley (Eds.), *The sustainable urban development reader*. London & New York: Routledge.

Goldstein, S. (1990). Economic and social impact considerations in highway programs. *Urban Transportation Systems*, p. 3.

Gordon, P., & Richardson, H. (1997). Are compact cities a desirable planning goal? *Journal of the American Planning Association, 17*(5).

Gordon, P., & Richardson, H. (2000). *Critiquing sprawl's critics.* Retrieved from Washington: https://object.cato.org/pubs/pas/pa365.pdf

Gordon, P., & Wong, H. L. (1999). The costs of urban sprawl: Some new evidence. *Environment and Planning A, 17*(5).

GSG. (2005). *Smart growth for brownfield redevelopment.* Retrieved from Chicago.

Gyourko, J. E., & Rybczynski, W. (2000). Financing new urbanism projects: Obstacles and solutions. *Housing Policy Debate, 11*(3).

Hagerty, R. (2012). *Building up while building out: Residential infill and smart growth development in metro Atlanta.* Retrieved from https://smartech.gatech.edu/handle/1853/43466

Hall, P., & Pfeiffer, U. (2013). *Urban future 21: A global agenda for twenty-first century cities.* London: Routledge.

Handy, S. (2005). Smart growth and the transportation-land use connection: What does the research tell us? *International Regional Science Review, 28*(2).

Harris, J. M. (2003). Sustainability and sustainable development. *International Society for Ecological Economics, 1*(1), 1–12.

Harris, M. (2004). What is the state of smart growth today, getting smart. *Newsletter of the Smart Growth Network.*

Hess, E. D. (2016). *Grow to greatness: Smart growth for private businesses.* Stanford: Stanford University Press.

Hodges, H. K. (2009). *City of south lake Tahoe subdivision ordinance: An opportunity for smart growth, sustainability and application streamlining.* (Master of City and Regional Planning), California Polytechnic State University, San Luis Obispo.

Holloway, J. E., & Guy, D. C. (2000). Smart growth and limits on government powers: Effecting nature, markets and the quality of life under the takings and other provisions. *Journal of Environmental Assessment Policy and Management, 9.*

Home Builders Association. (2002). HBA of NKY community development policies. Retrieved from www.hbanky.com/GovtAffairs/CommDevPol.asp

HosseinZadeh Dalir, K., & Hoshyar, H. (2005). Modernism and its influence on Iranian architecture and urbanism (In Farsi). *Geography and Planning, 19*(11), 207–222.

HosseinZadeh Delir, K, & Safari, F. (2012). The effect of smart planning on urban spatial planning. *Journal of Geography and Urban Development, (1)* (Spring and Summer).

Hwang, H.-M., & Foster, G. D. (2006). Characterization of polycyclic aromatic hydrocarbons in urban stormwater runoff flowing into the tidal Anacostia River, Washington, DC, USA. *Environmental Pollution, 140*(3).

Jacobs, J. (1961). *The death and life of great American cities.* New York: Random House.

Jensen, J. (2013). *Toward a new paradigm of sustainable development.* New York: CSIS.

Johnson, M. P. (2001). Environmental impacts of urban sprawl: A survey of the literature and proposed research agenda. *Environment and Planning A, 33.*

Kackar, A., & Preuss, I. (2003). *Creating great neighborhoods: Density in your community.* Local Government Commission in cooperation with the Environmental Protection Agency.

Keating, A. D. (2008). *Chicago neighborhoods and suburbs: A historical guide.* University of Chicago Press.

Kennedy, C., Steinberger, J., Gasson, B., Hansen, Y., Hillman, T., Havranek, M., . . . Mendez, G. V. (2009). *Greenhouse gas emissions from global cities.* Washington: ACS Publications.

Knaap, G., & Talen, E. (2005). New urbanism and smart growth: A few words from the academy. *International Regional Science Review, 28*(2).

Krieger, A. (1991). *Andres Duany and Elizabeth plater-zyberk: Towns and town-making principles.* New York: Harvard university graduate school Design.

Labi, S., Gkritza, K., Sinha, K. C., & Mannering, F. L. (2008). Influence of highway construction projects on economic development: An empirical assessment. *The Annals of Regional Science, 42*(3).

Leitman, J. (1993). Rapid urban environmental assessment: Lessons from cities in the developing world. *Urban Management Programme, 1.*

Lewis, R. (2011). *Do smart growth instruments in Maryland make a difference?* (Doctor of Philosophy), University of Maryland, USA.

Litman, T. (2004). *Understanding smart growth savings: What we know about public infrastructure and service cost savings, and how they are misrepresented by critics.* Retrieved from USA.

Litman, T. (2005). *Evaluating criticism of smart growth.* Victoria, CA: Victoria Transport Policy Institute.

Litman, T. (2010). *Affordable-accessible housing in a dynamic city: Why and how to support development of more affordable housing in accessible locations.* UDA: Victoria Transport Policy Institute.

Litman, T. (2015a). *Evaluating criticism of smart growth.* Victoria, CA: Victoria Transport Policy Institute.

Litman, T. (2015b). Urban sprawl costs the American economy more than $1 trillion annually: Smart growth policies may be the answer. *American Politics and Policy Blog.*

Litman, T. (2016). *Understanding smart growth savings evaluating economic savings and benefits of compact development, and how they are misrepresented by critics.* Retrieved from USA.

Local Government Commission. (1991). *Ahwahnee principles for resource-efficient communities.* Sacramento, CA: LGC.

Lucy, W. (2002). *Danger in exurbia: Outer suburbs more dangerous than cities.* Virginia: University of Virginia.

Lynch, K. (1981). *A theory of good city form.* Cambridge: MIT Press.

Hankey, S., & Marshall, J. D. (2010). Impacts of urban form on future US passenger-vehicle greenhouse gas emissions. *Energy Policy, 38*(9).

McCann, B. A., & Ewing, R. (2003). *Measuring the health effects of sprawl: A national analysis of physical activity, obesity and chronic disease.* Retrieved from http://smartgrowth.umd.edu/assets/ewingmccann_2003.pdf.

McElfish, J. M. (2007). *Ten things wrong with sprawl.* Washington, DC: Environmental Law Institute.

McManus, P. (2005). *Vortex cities to sustainable cities: Australia's urban challenge.* Sydney: University of New South Wales Press.

Miles, R., & Song, Y. (2009). "Good" neighborhoods in Portland, Oregon: Focus on both social and physical environments. *Journal of Urban Affairs, 31*(4), 491–509.

Miller, J. S., & Hoel, L. A. (2002). The "smart growth" debate: Best practices for urban transportation planning. *Socio-Economic Planning Sciences, 36*(1).

Mirowsky, J., & Ross, C. E. (2015). Education, health, and the default American lifestyle. *Journal of Health and Social Behavior, 56*(3), 297–306.

Molinaro, J. (2003). *Higher density, mixed-use, walkable neighborhoods – do interested customers exist?* National Association of Realtors.

Nahlik, M. (2014). *Transit-oriented smart growth can reduce life-cycle environmental impacts and household costs in Los Angeles.* (PhD dissertation), Arizona State University, Arizona.

Nazarnia, N., Schwick, C., & Jaegera, J. A. G. (2016). Accelerated urban sprawl in Montreal, Quebec City, and Zurich: Investigating the differences using time series 1951–2011. *Ecological Indicators, 60.*

Ogola, P. F. A. (2007). Environmental impact assessment general procedures. *KenGen,* 2–17.

Papa, E., & Bertolini, L. (2015). Accessibility and transit-oriented development in European metropolitan areas. *Journal of Transport Geography, 47,* 70–83.

Parfrey, E. (2003). What is smart growth? Retrieved from www.sierraclub.org/sprawl/com munity/smartgrowth.asp

Peltenburg, M., Davidson, F., Wakely, P., & Teerlink, H. (1996). *Building capacity for better cities.* Rotterdam: Institute for Housing and Urban Development Studies.

Porter, D. R. (2002). *Making smart growth work.* Washington, DC: Urban Land Institute.

Preuss, I., & Vemuri, A. W. (2004). "Smart growth" and dynamic modeling: Implications for quality of life in Montgomery County, Maryland. *Ecological Modelling, 171*(4).

Rahnama, M. R., & Abbas Zadeh, G. (2006). Comparative study of distribution/compression grading in metropolitan Sydney and Mashhad. *Geography and Regional Development Magazine.*

Reconnecting America's Center for Transit-Oriented Development. (2008). *Station area planning: How to make great transit-oriented places.* Retrieved from USA: www.Reconnect ingAmerica.com

Reese, I. (2011). *Altoona, PA: Researching smart growth principles in a shrinking city.* Pennsylvania: Pennsylvania State University.

Renne, J. L. (2005). *Transit-oriented development: Developing a strategy to measure success.* Transportation Research Board.

Royal Government of Bhutan. (2012). *Defining a new economic paradigm.* Retrieved from New York: https://sustainabledevelopment.un.org/index.php?page=view&type=400&nr=61 7&menu=35

Schlossberg, M., & Brown, N. (2004). Comparing transit-oriented development sites by walkability indicators. *Transportation Research Record: Journal of the Transportation Research Board,* (1887), 34–42.

Schuftan, C. (2003). The emerging sustainable development paradigm: A global forum on the cutting edge of progressive thinking. *The Fletcher Journal of International Development, XVIII.*

Seifuddini, F., & Shorjah, M. (2014). *Smart planning of land use and urban transportation, a dialectical look at urban space.* Tehran: Modiran – e- Emrooz Press.

Shafiea, F.A., Omara, D., & Karuppannan, S. (2013). Environmental health impact assessment and urban planning. *Procedia – Social and Behavioral Sciences, 85.*

Shapiro, J. M. (2005). *Smart cities: Quality of life, productivity, and the growth effects of human capital.* Retrieved from Cambridge.

Sheehan, M. O. M. (2001). *City limits: Putting the breaks on sprawl.* Retrieved from Washington: http://www.worldwatch.org/system/files/WP156.pdf

Simmons, D. (1999). *The problems with planning: A free-market guide to suburban development & "urban sprawl".* Competitive Enterprise Institute. Retrieved from https://cei.org/stud ies-issue-analysis/problems-planning-free-market-guide-suburban-development-urban-sprawl

Simons, R. (1998). *Turning brownfields in to greenbacks.* Retrieved from Washington DC.

Smart Growth Leadership Institute. (2004). *Smart growth is smart business.* Retrieved from USA: https://smartgrowthamerica.org/resources/smart-growth-is-smart-business/

Smart Growth Network. (2002). *Getting to smart growth: 100 policies for implementation.* Retrieved from www.smartgrowth.org.

Smart Growth Network. (2006). *This is smart growth.* Princeton: Princeton Junction.

Strafford Regional Planning Commission. (2011). *How to link environmental acts and transportation planning.* Retrieved from http://strafford.org/uploads/2009SRPCCommissionHa ndbook_final.pdf

Snyder, K., & Bird, L. (1998). *Paying the costs of sprawl: Using fair-share costing to control sprawl.* Washington, DC: The U.S. Department of Energy.

Song, Y. (2005). Smart growth and urban development pattern: A comparative study. *International Regional Science Review, 28*(2).

Squires, G. D. (2002). *Urban sprawl: Causes, consequences, & policy responses.* Washington, DC: The Urban Institute Press.

Sroka, R. (2016). TIF for that: Brownfield redevelopment financing in North America and Calgary's rivers district. *Cambridge Journal of Regions, Economy and Society, 9*(2), 391–404.

Staley, S. R. (2004). Urban planning, smart growth, and economic calculation: An Austrian critique and extension. *The Review of Austrian Economics, 17*(2).

Stevenson, M. (1995). *Social impact assessment of major roads.* Toronto: Hardy Stevenson and Associates

Strafford Regional Planning Commission. (2011). *How to link environmental acts and transportation planning.* Retrieved from http://strafford.org/uploads/2009SRPCCommissionHandbook_final.pdf

Surface Transportation Policy Project. (2003). *Measuring the health effects of sprawl.* Washington, DC: Smart Growth America.

Suzuki, H., Cervero, R., & Iuchi, K. (2013). *Transforming cities with transit: Transit and land-use integration for sustainable urban development.* The World Bank.

Talen, E., & Knaap, G. (2003). Legalizing smart growth an empirical study of land use regulation in Illinois. *Journal of Planning Education and Research, 22*(4).

The Real Estate Research Corporation. (1974). *The costs of sprawl.* Washington, DC: U.S. Government Printing Office.

Theart, A. (2007). *Smart growth: A sustainable solution for our cities?* (Master of Philosophy in Sustainable Development Planning), University of Stellenbosch, South Africa.

Trillo, C. (2013). Urban sprawl management, smart growth: Challenges from the implementation phase. *International Journal of Society Systems Science, 5*(3), 261.

Trust for Public Land. (1999). *The economic benefits of parks and open space: How land conservation helps communities grow smart and protect the bottom line.* Washington, DC: Trust for Public Land.

Tsai, Y. H. (2005). Quantifying urban form: Compactness versus ' sprawl'. *Urban Studies, 42*(1).

Tszmokawa, K., & Hoban, C. (1997). *Impacts on communities and their economic activity.* Washington, DC: The World Bank.

Turner, M. (2006). A simple theory of smart growth and sprawl. *Journal of Urban Economics, 61.*

Twaddell, H., & Emerine, D. (2007). *Best practices to enhance the transportation-land use connection in the rural United States* (Vol. 582). Washington, DC: Transportation Research Board.

U.S. Department of Agriculture, E. R. S. (2001). *Smart growth: Implications for agriculture in urban fringe areas.* Retrieved from www.ers.usda.gov/publications/AgOutlook/April2001/ao280.pdf

U.S. Department of Housing and Urban Development. (2003). Smart growth and livable communities resources. Retrieved from www.hud.gov/offices/cpd/economicdevelopment/programs/rc/resource/smartliv.cfm

U.S. Environmental Protection Agency. (2002). *Smart growth policy database glossary.* Washington, DC: Publication EPA.

U.S. Environmental Protection Agency. (2012a). *Smart growth and economic success: Benefits for real estate developers, investors, businesses, and local governments.* Washington, DC: Environmental Protection Agency.

U.S. Environmental Protection Agency. (2012b). *Smart growth and economic success: Strategies for local governments.* Retrieved from USA.

U.S. Environmental Protection Agency. (2012c). *Smart growth and equitable development.* Retrieved from USA.

Urban Land Institute. (2003). *What is smart growth?* Retrieved from http://smartgrowth.net/Home/sg_Home_Main_A.html

USEPA. (2011). *Essential smart growth fixes for urban and suburban zoning codes*. Retrieved from Washington, DC, www.epa.gov/smartgrowth.

Vanthillo, T., & Verhetsel, A. (2012). Paradigm change in regional policy: Towards smart specialisation. *Belgian Journal of Geography, 2* (Inaugural).

Vos, J. D., Acker, V. V., & Witlox, R. (2016). Urban sprawl: Neighbourhood dissatisfaction and urban preferences. Some evidence from Flanders. *Urban Geography, 37*(6), 839–862.

White, S. M., & McDaniel, J. B. (1999). *The zoning and real estate implications of transit-oriented development*. Washington, DC: National Research Council, Transportation Research Board.

Williams, C. J. (2003). Do smart growth policies invite regulatory takings challenges: A survey of smart growth and regulatory takings in the southeastern United States. *Alabama Law Review, 55,* 895.

Wu, J. (2006). Environmental amenities, urban sprawl, and community characteristics. *Journal of Environmental Economics and Management, 56*(2).

Yang, F. (2009). *If 'Smart' is 'Sustainable'? an analysis of smart growth policies and its successful practices*. (Master of Community and Regional Planning), Iowa State University, Iowa.

Zhang, L., Xu, W., & Li, M. (2009). *Co-evolution of transportation and land use: Modeling historical dependencies in land use and transportation decision-making* (No. OTREC-RR-09-07). Oregon Transportation Research and Education Consortium.

Zhao, P. (2010). Sustainable urban expansion and transportation in a growing megacity: Consequences of urban sprawl for mobility on the urban fringe of Beijing. *Habitat International, 34*(2), 236–243.

Ziegler, E. H. (2003). Urban sprawl, growth management and sustainable development in the united states: Thoughts on the sentimental quest for a new middle landscape. *Virginia Journal of Social Policy & the Law, 11,* 26.

2

SMART GROWTH VS. URBAN SPRAWL

Introduction

Human societies have been changing ever since their formation. In the same vein, in line with the growth of various aspects of human life, cities have also changed and expanded as the main settlements of modern humans. This expansion has occurred either due to an increase in urban population, leading to the expansion of urban area, or as a response to the needs of twentieth-century people, such as the need to live in nature or to avoid the crowds of urban centres. As population growth or the people's interest in living in nature is not negative in itself, urban growth and expansion of cities cannot be regarded as undesirable. However, what can be a negative factor in the expansion of cities is related to the process and the way in which growth takes place. Urban growth is an inevitable phenomenon caused by factors such as rapid population growth; the moment-by-moment development of knowledge, technology and information systems; change in urban economic structures, etc. Urban growth is negative if it happens in an unplanned, uncontrolled and inconsistent manner in such a way as to create scattered and single-use areas that exceed the capacity and needs of cities.[1] Urban planners use the term "sprawl" to refer to the negative side of horizontal urban growth.

Since the 1950s, with the development of interstate highway systems, the (American) pattern of residential development has become more dispersed. This development pattern has accelerated over the past 50 years, despite debates on the energy crisis and recession. Sprawl increases the cost of development in the suburbs and reduces the environmental factors needed for sustainable economic growth. It also results in high consumption of agricultural land, energy and natural resources. One of the concerns in such a situation is the opposition of the consequences of this development pattern to the long-term interests of cities and human settlements (Chen, 2008).

Although urban areas account for only 14 percent of the earth's surface, the loss of land as a result of urban sprawl cannot be underestimated, as the expansion of cities has greater environmental impacts, such as floods, landslides, changes in soil fertility, etc. than other types of land use do (Batisani & Yarnal, 2009).

The dominant pattern of urban growth after World War II has been scattered and mostly apart from the old city centres. This has caused cities to expand horizontally, leading to enormous consequences for the environment and economy of cities, as well as for the human community. These consequences include scattered urban development, destruction of agricultural land and gardens, evacuation and destruction of central old contexts, barriers to service provision due to the outbreak of urban sprawl, environmental problems and pollution caused by excessive use of vehicles, as well as increased city limits and the destruction of natural centre (Clawson & Hall, 1973).

What is urban sprawl?

The coinage of the term "sprawl" in the urban literature

The term "urban sprawl" was first used in 1958 by William H. Whyte (a sociologist) in an article entitled "The Exploding Metropolis", in *Fortune* magazine. White's explicit sense of the word "sprawl" is the physical growth of cities towards the

FIGURE 2.1　Low-density housing in U.S. suburbs

Source: Jensen, 2005

outskirts caused by factors such as building towns, the development of transport networks and, ultimately, urban management policies (Whyte, 1993). Since then, planners have used this term to refer to a type of urban development which has adverse social, environmental and economic effects. Horizontal urban growth is synonymous with the spatial expansion of urban areas to undeveloped areas, which often leads to issues like suburbanism. But sprawl has an inherent negative meaning, so that according to Ewing, it is a relatively new model in human settlements that is accompanied by a random gathering of residential units (low density; see Figure 2.1) and strip-shaped business developments. And this phenomenon derives from the widespread use of cars in urban life (Ewing, Hamidi, & Grace, 2016)

A brief overview of urban sprawl definitions

Given the various definitions suggested for urban sprawl, Table 2.1 summarizes only some of these definitions and briefly describes the dimensions and characteristics of this urban development model.

TABLE 2.1 Definitions of sprawl suggested by researchers, scholars and accredited international institutions

Definition	Source
Urban sprawl refers to scattered, unplanned, self-reliant and uncontrolled expansion of cities away from the dense centres of towns and villages along the highways in the suburbs.	(Menon, 2004)
Sprawl is the shift of development away from the centralized and fully developed areas in the centre of a city and its peripheral areas towards undeveloped or rarely developed scattered areas.	(Speir & Stephenson, 2002)
Sprawl is a special type of suburban development with characteristic features of low-density residential and non-residential settlements, the dominance of car-driven transportation, horizontal outward growth of urban uses and land use separation.	(USHUD, 1999)
Sprawl means excessive land consumption, constant development, unconnected development and inefficient use of land.	(Peiser, 2001)
Extremely low-density development outside urban centres, usually on undeveloped land.	(Snyder & Bird, 1998)
Sprawl in essence occurs around cities, where land is not expensive and, as far as possible, its development patterns are spatially open and free from rules and regulations in comparison with the central city.	(Tabibian & Asadi, 2009)
Physical incoherence in the spatial expansion of cities leads to sprawl, the important adverse effects of which include an unplanned rise in social service and public transport costs, inadequate access to public open spaces, etc.	(Clawson, 1962)

Definition	Source
The widespread, dispersed and unconnected development can be dubbed urban sprawl.	(Ludlow, 2006)
Expansion of cities in the suburbs toward villages and/or along highways, or unplanned and uncontrolled expansion across the city.	(Hadley, 2000)
Sprawl means the distribution or destruction of landscapes and ecosystems through thin and scattered development of urban settlements out of the developed regions.	(Urban Sprawl Inc, 2015)
Urban sprawl refers to the greedy use of land, uneven and constant development, scattered development and ineffective use of land.	(Couch & Leontidou, 2007)
Sprawl refers to decentralization of populations and employment from urban centres towards the suburbs. This process can also be called decentralization or suburbanization.	(Mills, 2003)
The spread of new construction in fragmented lots, separated from each other by empty barren land.	(Ottensmann, 1977)
Sprawl is dispersed development with a tendency to build outside urban centres along highways. It has features such as excessive land consumption, lower density compared to old centres, high vehicle dependency, differentiation of open spaces, large spaces between built lines, separation of uses in different areas, single-storey commercial buildings and lack of civil and public spaces.	(Carruthers & Ulfarsson, 2002)
Sprawl is a special form of urban development characterized by low density, dispersed development and imposition of harmful social and environmental impacts.	(Poelmans & Rompaey, 2009)
Sprawl is an unplanned, uncontrolled, uncoordinated and single-use development. It does not encourage mixed uses and does not have any functional relationship with its surrounding uses. It is thus manifested as a kind of linear or striped, dispersed and isolated development.	(Nozzi, 2003)
Sprawl is a process where development and expansion are faster than population growth. Urban sprawl has four dimensions: a) a population that is widely distributed to low-density developed areas; b) houses, stores, and workplaces are separated from one another; c) it has a network of roads with large-scale blocks and poor access; and d) it has a spatial structure in which the centres are not defined.	(Seifuddini & Shorjah, 2014)
Sprawl refers to random low-density development in a widespread area where single-family dwelling is the dominant residential form. The obvious consequences of this urban pattern include the social isolation of individuals, global warming due to personal car contaminants, flooding and erosion due to the growing paving land, the decline of small farms, the destruction of wildlife and the destruction of the balance of nature.	(Hughes & Seneca, 2004)

(Continued)

TABLE 2.1 (Continued)

Definition	Source
Sprawl is development in areas far from the traditional urban centres where the use of cars is a priority in place of other forms of transportation. Therefore, such regions have the highest rate of distribution, leading to the decline of the traditional city centre. Almost all residents in these areas use cars for commuting.	(Lewyn, 2000)
Sprawl is a scattered distribution of urban settlements in rural areas and landscapes. In fact, it happens when two neighbouring cities penetrate the surrounding rural lands under the effect of development and create an area in which some parts of the farms remain intact while the rest are developed.	(Harvey & Clark, 1965)
Sprawl is an inappropriate development pattern characterized by a highly wasteful pattern of land development, because it is associated with a tendency towards unnecessary consumption of a huge amount of natural resources. In this pattern of physical expansion, there is a great need to invest heavily in building public infrastructure and services, which exhausts energy and human resources by forcing them to travel long distances.	(H.C. Planning Consultants, 1999)
Sprawl is a form of urban growth, which affects the social, ecological and aesthetic functions of the city, despite having short-term benefits for individuals. Better to say, it is an inefficient, wasteful urban growth city into the suburbs.	(Hess, 2001)
The term urban sprawl is referred to as a growth that has the following characteristics: a) low-density population and construction; b) disconnected urban fabric or strip development; c) automobile dependency for daily trips; and d) the loss of agricultural land through differentiation.	(Chen, 2008)
"Urban sprawl" is widely defined by many urban planners as the "creeping progress" of residential and non-residential development in rural and underdeveloped areas, often characterized by leaping and isolated development and high land consumption. They also emphasize that the meaning of the term is context dependent, so that it varies from one place to another and from one culture to another.	(Burchell, 2000)
In urban environments characterized by sprawl, the population is widespread in low-density residential constructions; shops and workplaces are rigidly segregated; and there is a lack of vivacious centres, such as strong city or town centres. Urban sprawl is the result of statistical facts and human conditions. Statistically, sprawl results from demographic factors such as population growth and spatial factors, including restrictions on the city's borders. When urban centres reach their final capacity, the surplus must to settle elsewhere (i.e. suburbs).	(McCann & Ewing, 2003)

Definition	Source
Urban sprawl is discrete development out of dense city and rural centres along highways, with the main characteristics including high land consumption, lower-than-average density compared to old centres, automobile dependency, segmented open spaces, large distances among built surfaces, separation of uses in different areas, single-story commercial buildings and lack of public space and local centres.	(Vermont Forum on Sprawl, 1999)
In urban sprawl, natural resources and fertile agricultural lands disappear in favour of urban development, ignoring the opportunity to use land and buildings within cities, towns and villages to meet urban growth needs.	(Ewing, Kostyack, Chen, Stein, & Ernst, 2013))
Urban sprawl is referred to as "scattered growth in the outskirts of cities" and includes the dissemination of the built context of a city to agricultural lands around the urban area. Residents of neighbourhood units emerged as a result of urban sprawl, prefer living in large single-family homes and commute to workplaces or educational centres with private cars. Urban planners in this type of development further point to the qualitative disadvantage of sprawl, such as the lack of transportation options and of suitable neighbourhoods for walking. On the other hand, environmental advocates stress the high amount of land consumed through urban sprawl.	(U.S. Department of Agriculture, 2001)
Spread is defined as a land use pattern which is characterized by sporadic and random development. Although sprawl is typically recognized as a phenomenon related to urban areas, its problems can spread to suburban, rural and even interstate areas.	(Gray, 2005)
[Sprawl] is a special kind of urban development characterized by low density of both residential and non-residential uses, dominant use of private cars, unlimited outward expansion of new developments and distinction between land uses based on the type of activity.	(U.S. Department of Housing and Urban Development, 2003),
[Sprawl] is an irresponsible development that takes away tax resources from our societies and eliminates farmlands and open spaces.	(Smart Growth Network, 2006)
Sprawl is the outbreak of external city borders caused by lack of land use planning. Development in this pattern is isolated, scattered and outward from the city centres. Sprawled cities, in contrast to the densely populated cities, are full of empty spaces indicating inefficiencies in development and manifesting uncontrolled growth.	(Environmental Protection Agency, 2013(

Causes of urban sprawl

As Baum (2004) states, sprawl is the product of "suburban pulls" and "urban pushes". The pull factor is aggravated by the dream of living in a less dense rural setting under the pressure of anti-urban thoughts against living in intensive urban centres (Baum, 2004). Pull and push factors may be influenced by a variety of other variables, such as market dynamics, demographic factors and housing supply and demand. However, a person's preference is completely subjective and rooted in his or her personal benefits (in the short term). In this way, raising citizens' awareness regarding the need for a long-term approach to urban planning is essential in order to achieve sustainable patterns of development in the future. Bengston, Fletcher and Nelson (2004) believe that urban sprawl benefits new residents, investors and other stakeholders separately. However, sprawl is increasingly seen as an important and growing concern that imposes a wide range of social and environmental costs – especially in the area of rising energy prices. According to Robinson, Newell and Marzluff (2005), awareness of the costs of sprawl has pushed policymakers around the world to create various regulations and incentives to reduce this phenomenon.

According to Gordon and Richardson (1997), "some scholars consider contemporary development patterns [to be] a reflection of the 'invisible hand of the market'". However, according to Wang (2004), we should admit that besides the market, other factors contribute to urban sprawl. Some of these factors are mentioned here.

a) **Metropolitan population growth**. When the urban population exceeds a certain threshold, an external expansion will inevitably occur in response to the new need for development. In the last half century, the metropolitan population has grown from 85 million to 213 million, indicating a 150 percent increase in the population and 70 percent growth in the suburbs (Stephens & Wikstrom, 2000).

b) **Abundance of land**. In the United States, the abundance and availability of land are responsible for the development of residential and non-residential areas in high amounts. Nowadays, in response to new needs, changes in urban areas are inevitable, imposing direct surplus costs for the purchase of buildings and for demolition, preparation and reconstruction. In many cases, construction regulations in existing urban contexts pose many obstacles to construction. Hence, if there is enough vacant land in the suburbs around cities, they will quickly be seized for development and to satisfy these new needs (United Stated Department of Agriculture, 2000)

c) **Industrial revolution and developing new technologies.** The industrial revolution, which brought new technologies as its by-product, is a milestone in the history of urbanization, as many authors considered developed technologies at that time a major factor for urban sprawl. In this regard, Gillham (2002) writes: "The contemporary suburban and urban sprawl are the result of the industrial revolution. The rapid growth and changes in the cities as a result of the industrial revolution have brought about modern suburbs. Factories, mass production and new forms of transportation and communications have guided us to what we are heading today".

d) **Decentralization of employment**. Certainly, in most cases, the costs of land purchases and development outside urban areas are less than those inside urban areas. On the other hand, modern high-speed systems also reduce transportation costs, increasingly pushing factories and industries toward the suburbs, where they have easier access to highways. There is also a well-trained and skil-ful workforce living in the suburbs for various reasons, such as benefiting from a desirable atmosphere with privacy and pleasantness. In these conditions, the scattering of employment places is the cause rather than the consequence of residential suburbanization. Increasing job opportunities also encourages more people to migrate to suburban areas (Carruthers & Ulfarsson, 2002).

e) **Housing priorities**. Separate and safe houses, suburban neighbourhoods with grasslands in the background, one or two cars for commuting, good and close public schools . . . all are typical of the urban life known as the "American Dream". In fact, this is the dream of all people. When urban residents become richer over time, they can achieve larger houses in urban areas with lower prices. In addition, thanks to technology advances, the American Dream does not mean an isolated rural life. Rather, it makes it possible to have an urban life in a rural setting (The Real Estate Research Corporation, 1974(.

f) **Destruction of city centres**. City centres used to be good places for people to live. However, with the growth in population and the economy, city centres have gradually declined due to issues such as traffic congestion, environmental degradation, stagnant housing, poor-quality public schools, increased crime, lack of access to open spaces and the destruction of infrastructure in such cen-tres. In search of a better life, the middle and upper classes are migrating outside the urban centres. As the number of jobs in suburban areas grows, more of the workforce emigrates from urban centres. When middle-class people escape from urban centres, the quality of life in such cores declines due to financial deficiencies, and these urban areas take on an undesirable appearance. Also, with the migration of people to places outside the city, taxes collected in cities and towns are used to develop the services and infrastructure of the suburbs. In such a situation, the business centres of cities deteriorate due to the lack of financial revenues.

g) **Transportation progress**. Sprawl is a transport-based phenomenon. Growth happens in most cities on a regular basis, but the push for compatibility with technology, road quality and access to transit accelerates the process. More than 2,000 years ago, walking and chariots were the main pillars of transporta-tion that caused the formation of densely populated cities due to their inher-ent limits in speed and distance travelled. The Roman roads had a maximum length of five kilometres and reached eight kilometres with the borders around them. Later on, towns expanded when trams and railroads were built and travel times became longer. As a result, private cars increased sprawl and accelerated the development of road systems. In fact, a shift from walking to urban tram and then the private and public transport systems transformed human lifestyles. Personal cars also made trips easier, more convenient and more practical. In addition, after World War II, zoning regulations differentiated between differ-ent types of land use, such as workplace, shopping place and accommodations, and people were forced to resort to their own cars to travel to these separate destinations (Black, 1991).

h) **Public policies.** The urban sprawl pattern is not just an individual or corporate issue regarding spatial deployment preferences. The choices of people in this regard are largely influenced by numerous government policies, including tax policies, depreciation funds, zoning regulations and implicit donations, grants for suburban development and discouragement of rehabilitation efforts in inner-city and suburban areas (Office of Technology Assessment, 1995)

i) **Land value.** The price of agricultural land is quite lower than plots, and this is an important factor causing urban sprawl (Uhel, 2006). Low land prices cause a heavy demand by sectors such as industry, trade and construction. Preference of these sectors for agricultural land instead of plots causes both urban sprawl and inefficient use of resources due to non-agricultural use of agricultural land (Ludlow, 2006).

Aspects of urban sprawl

Urban sprawl and economy

Mokhtari, Hosseinzadeh and Alizadeh (2013) argue that

> dispersed growth is formed due to transformation of cities' economic foundation, land exchange, inconsiderate urban planning policies, and sudden decisions for urban development, and it is itself the cause of adverse ecological, social and economic consequences, making cities confronted with a complicated problem.

It could be argued that if land is scarce and expensive, there will be less of a market for sprawl development. For example, in the Netherlands more compact design and higher density have become a norm out of necessity. The urban sprawl paradigm removes the economic and cultural life of cities and consumes more fertile land and fuel for transportation. On the one hand, the world's population is constantly rising. Therefore, in cities and towns around the world, older urban centres are overlooked and abandoned. On the other hand, development has been brought to the green lands, and the demand for personal cars to travel among these scattered destinations is growing steadily (Halleux, Marcinczak, & Krabben, 2012). Low-density development in the suburbs leads to the construction of separate residential units, and as a result of such a demand, it is drawn to empty lands or agricultural fields in the countryside.

Urban sprawl and land use patterns

Land use patterns form one of the elements used to define sprawl. Negative aspects of urban sprawl include the high consumption of agricultural land and land with environmental significance, reliance on personal cars for transportation and, in general, a lack of integrated land use planning.

Separation and differentiation of land uses is another characteristic of sprawl. In dispersed areas, residential units are often distant from the city centres, shopping malls and offices. Due to this separation of uses, every city trip requires the use of cars. More traditional patterns of development tend to have mixed uses, especially with housing units placed near shopping malls, offices or in floors over shops. Hence, the degree of use mixing can be considered an important variable in the description of urban sprawl (Banai & DePriest, 2014). Ewing, Pendall and Chen (2002) argue that

> the mixed use variable reflects the balance between occupations and population, the distribution of uses within local communities, and the availability of residential units to shopping centres and schools. Higher densities allow for more mixed use; therefore, the mixing factor is relatively related to the factor of density. However, the evaluations show that the mixing factor is clearly distinct from the density factor.
>
> *(Ewing et al., 2002)*

Urban sprawl and density

Many definitions of urban sprawl use a low-density criterion to determine scattering; however, this concept is not quantitative and does not have a specific standard. On the contrary, this criterion is relative and based on the cultural conditions of each country. For example, in the United States, low density is usually expressed as development with 2 to 4 residential units per acre (10 to 15 residential units per hectare), whereas in England, low density refers to 8 to 12 residential units per acre (20 to 30 residential units per hectare).

Residential density is one of the most important indicators of sprawl. Separation and division of urban suburbs are the characteristic features of sprawl that hinder the provision of services and the availability of facilities to residents and make it difficult to access urban centres or transportation options. However, higher density does not necessarily mean high-rise building (Banai & DePriest, 2014). Ewing et al. (2002) argue that the type of density that encourages smart growth can range from 30 to 35 residential units per hectare and be in single-family neighbourhoods. Such densities allow nearby neighbourhoods to benefit from available shopping malls, small neighbourhood schools and more public transportation services. Residential density is the quantitative measure of the amount of land used by each person and shows the degree of sprawl or compaction of housing (Ewing et al., 2002)

Urban sprawl and morphological patterns

According to Hess (2001), "the term 'urban sprawl' dates back to the mid-twentieth century, when the expansion of urban spaces flourished in the United States as a result of the excessive use of personal vehicles and the development of highway system."

One of the first criticisms heralded at the form of development that results in urban sprawl was posed by Jane Jacobs in 1961 in an article "The Death and Life of Great American Cities". Jacobs believes that, unlike life in the suburbs and away from the community, the combination of small-scale diverse uses creates lively and successful neighbourhoods (Jacobs, 1961). Urban sprawl is associated with automotive transportation, which affects the social health of families and society: "Life that should be spent on enjoying an enriched society is increasingly becoming a lonely life behind the steering wheels of cars" (Duany, 2002).

The passageway network can be dense or dispersed, interconnected or interrupted. Consequently, the blocks that are created by the streets can be short and small or tall and large. The crowded arteries fed from dead-end residential streets form an example of a sprawling passageway network. These arteries create very large blocks that focus vehicle traffic into a limited number of routes and make access via public transport, hiking or cycling impossible. Compact development generally consists of a network of interconnected streets with shorter blocks that provide more access via a wider range of options for drivers, pedestrians and cyclists (Banai & DePriest, 2014)

From Hess's point of view, the word "sprawl" is an adjective describing a kind of urban growth pattern. As a verb, it describes its formation process, and as a noun, it defines the form of urban land.

When defining the concept of sprawl, the Sierra Club's research states that sprawl is fast and scattered growth throughout the motherland areas and even small towns that – in some cases – have been drawn up to the rural areas or the borders of villages. Sprawl is a low-density development outside the service and employment centres of the city that has led to the separation of houses from shopping, work, recreational and educational areas (Sierra Club, 2001). In the book *Urban Sprawl*, Squires defines sprawl as a pattern of metropolitan and urban growth that reflects low-density, automobile dependent, exclusionary new development towards natural areas around a city (Squires, 2002).

In explaining the concept of dispersion, Robert W. Wassmer (2002) writes:

> According to some researchers, horizontal distribution is in the form of scattered and discontinuous development and is associated with expansion to areas beyond the urban boundaries and suburbs, emphasizing the dominance of personal cars in transportation.

Also, according to Siembab (2005), "scattering is a less dense urban growth caused by a wide range of functional areas such as administrative parks, small business centres, and residential units."

According to Chin (2002), "there are various types of urban forms called 'urban scattering'." They include contiguous suburban growth, linear development (Figure 2.2), leapfrog development (Figure 2.3) and scattered development (Figure 2.4). In terms of urban form, scattering differs from the optimal and ideal condition of a compact city, with high density, centralized development

FIGURE 2.2 Linear development in urban sprawl in Milton, Ontario

Source: Simon, 2009

FIGURE 2.3 An example of a leapfrog development in Las Vegas

Source: Received from Google Map in 2018

FIGURE 2.4 Suburban development in a northeastern section of Colorado Springs, Colorado

Source: Shankbone, 2008

and mixed spatial functions; however, what should be considered in scattering is a chain (or spectrum) of the dense development toward completely scattered developments. Dispersion is actually a degree of this spectrum and is not a complete and absolute form.

Although at the end of the compression and density spectrum, there is a growth in urban suburbs, this growth is seen as a continuous expansion of existing developments from the core. Leapfrog or scattered development is at the other end of the spectrum. This form represents a discontinuous development far from the city's old core in the form of construction in scattered areas and vacant lands. In the current literature, this form is commonly known as "scattering". Analysts like Ewing (1997) draw a distinction between "leapfrog" and "scattered" development and consider leapfrog development to be a single-core city, whereas scattered development can be multi-core. An important problem in the definitions of urban dispersion is that different types of development, with a distinction between the contiguous growth of suburbs and scattered development, both are classified as "scattering", but their shapes and effects are very different. Therefore, it is better not to consider scattering an absolute and complete urban form, but rather a chain and a range from compressed to completely scattered development (Ewing, 1997).

The consequences of urban sprawl

Urban sprawl is associated with numerous environmental, social, transportation, economic and public health problems. These problems include the increase of unused land, increased share of open spaces, reduced population density, disrupted urban segments and social isolation. Jackson (1985) states:

> Although scattered growth is generally recognized by low population densities, low traffic, car dependency and large residential areas in suburban agricultural lands, its main characteristic is the high environmental and social costs of such growth.

However, Malizia and Exline (2000) maintain that living in single-family homes in suburban areas is preferable to many citizens (Malizia & Exline, 2000).

In general, the term "sprawl" has more negative connotations due to its environmental impacts and effects on people's health. Norman, MacLean, and Kennedy (2006) argue that residents of neighbourhoods that emerged as a result of urban sprawl generate more pollution per person and are faced with more commuting problems (Norman et al., 2006).

Ewing (1997) suggests that

> scattering is a sign of unexplained growth without planning in the city as a result of the separation of land uses, elimination of open spaces and degradation of the city core and the car-dependent transportation system, which is associated with increased traffic, fuel consumption and air pollution.

Gordon and Richardson (1997) also point out that "urban dispersion is an inevitable result of market priority in providing low-density residential areas and in preferring personal vehicles over public transport" (Gordon & Richardson, 1997). In this regard, Newman and Kenworthy (1999) argue that "urban scattering is not only a result of low densities, but also a result of scattered urban development". Use of cars is the first force that make cities to increase the use of land, energy, water, traffic and water pollution, and the resulting economic costs lead to a scattered infrastructure, direct and indirect transportation costs (road accidents, pollution, etc.), along with the loss of public resources (Newman & Kenworthy, 1999).

According to Peiser (2001), "urban scattering has obvious characteristics, which have been roughly one of the most pressing factors for American cities". Ewing (1997) also argues that

> scattering is a kind of waste of urban housing characterized by low uniform densities, often heterogeneous and spreading over the boundaries of urban areas at a high rate. Scattering in this process generally invades the primary agricultural areas and land resources, through which the land develops in a fragmentary manner, with many spaces left vacant or assigned to low-performance uses.

The city's scattered areas totally rely on cars for access to resources and facilities. Regarding scattering, Gillham (2002) writes:

> Scattering moves away from existing settlements as node (scattered) patterns. In many cases, distant development occurs first within several kilometers of the central employment segment, and then they are connected through inclusive development.

Therefore, urban sprawl appears to reduce population density, so that during this development, the city destroys agricultural lands and farms. Sprawl is manifested in a segmented, sporadic appearance and a discontinuous and gradual development.

Sprawl is associated with an increase in distances travelled, as well as the time for and number of journeys, and we can say that many of the negative consequences of this kind of urban growth are related to the issue of transportation.

According to Ewing (1997):

> In general, scattering disrupts the quality of city access in two ways: first, it causes poor access to residential land uses, because residents are often away from work, shopping and leisure options, and second, these land uses are separate from each other.

Table 2.2 shows some consequences of sprawl in cities and their suburbs.

TABLE 2.2 The consequences of sprawl in cities and their suburbs

Environmental degradation	Sprawl causes damage to the environment through the destruction of ecosystems, differentiation of forests, reduced air quality and biodiversity and threats to wildlife.
Agricultural land degradation	Sprawl is a serious threat to agricultural land and is responsible for its destruction. It also includes eliminating the identity of rural areas and the dominant modes of living therein.
Elimination of open spaces	Sprawl largely results in the destruction of open spaces and is considered to be a threat to them. Open spaces are distinct from agricultural lands and include a variety of vacant lands in urban areas such as playgrounds, forests, parks and wetlands.
Traffic problems	Sprawl plays an important role in traffic jams and prolonged travel time, contributing to air pollution and threats to the physical and mental health of citizens.
City centre decline	Sprawl plays a major role in the decline of major city centres. It causes a shift of public and private budgets from the renovation and rehabilitation of urban centres to the suburbs.
Elimination of the concept of community	Sprawl destroys the sense of belonging to a community and place and causes social isolation.

Lack of historical sites	Sprawl does not value historic and cultural sites such as historic buildings, old urban centres and memorable places, but is often considered a serious threat to them.
Financial consequences	Sprawl can increase the financial cost (both capital and recurrent) of providing urban services ahead of the compact city. Also, depending on personal vehicles for most of society's daily transportation can increase demands for fuel and increase the cost of living.

Source: Adopted from: (Bengston et al., 2004; Ibata, 2000; Parisi, 1998)

The consequences of urban sprawl based on impact location

Urban sprawl can have negative impacts on the countryside and cities. These are summarized as follows (Lewyn, 2000):

a)   **Effects on suburbs.** The rapid development and horizontal growth of cities have many negative effects on the suburbs. The expansion of residential units, linear shopping centres, the disappearance of office parks and the emergence of skyscrapers and apartment buildings are examples of these effects. The green suburban areas become grey after undergoing urban development. The inhabitants of these areas (unlike those of city centres) do not have walking access to public facilities (schools, shops, etc.) and have to use private cars. In such a situation, the suburbs will be worse off in terms of traffic. In addition, time spent to reach the workplace, bring children to schools and do many other things using cars will deprive suburban residents of leisure time. Ultimately, this sprawl and dependency on cars prevent daily interactions between neighbours and eliminate the sense of community that previously existed between residents of adjacent neighbourhoods.

b)   **Effects on cities.** The sprawl paradigm also affects a city's residents. Because in some cases rentals for residential units in the suburbs is higher than those in the city and ownership of cars is becoming more and more important for survival in the suburbs, poor people are deprived of the opportunity to live in the suburbs and can only stay in the city centres. Therefore, the average income (and subsequent tax revenues) of the city is reduced. Schools and urban services suffer losses, and buildings in the city are left empty and worn out. The moving of jobs to the suburbs also takes these opportunities away from city centres, and low-income people cannot afford a car to travel to these new locations.

Types of consequences of urban sprawl

In general, the consequences of urban sprawl can be classified into three types: environmental, social and economic.

a) **Environmental consequences**

According to the European Environment Agency (2006), the problem of the rapid consumption of rare earth resources is quite visible in the widespread growth of cities beyond their borders. Urban sprawl has led to an increased demand for raw materials and their transportation, especially in distant locations. Changes in land use alter the characteristics of water and land areas, which in turn leads to a change in the interaction between land and water. This goes so far that most of the watersheds affected by urban sprawl are faced with hydrological disturbances. The next consequence of increased land use and reduced population density due to urban sprawl is the increasing consumption of energy. Generally, congested urban developments with higher population densities are more efficient in terms of energy consumption. Increased use of personal transportation, which is accompanied by high energy consumption, leads to increased carbon dioxide emissions in the atmosphere. The link between population density and carbon dioxide emission shows that greenhouse gas emissions increase when urban density is reduced. Although several factors, such as the level of industrial activity and local climate conditions, can influence the degree of carbon dioxide production in different urban areas, automobile dependency in dispersed cities is an important factor in increasing greenhouse gas emissions. Sprawl and the consequent growth in urban transport and greenhouse gas emissions have had a major impact on global warming and climate change. It is expected that climatic conditions will become harsher and coastal and river floods will increase in the coming years. The dangers arising from the continued development of these areas in the form of climate change and the occurrence of large floods are quite evident.

b) **Economic consequences**

Low-density development in the suburbs requires more facilities and infrastructure. The number of roads and urban facilities sufficient for a certain number of people in the city will not be sufficient for the same number in the suburbs because they are more scattered. Creating more infrastructure will cost more and will have a negative impact on the economy. Also, from an economic perspective, sprawl imposes higher commuting costs on households by increasing home-to-work distances and makes access to jobs and businesses in scattered urban areas via personal automobile transportation systems unjustifiable.

Also, urban sprawl is a serious obstacle to the development of public transit systems and other alternative forms of transportation. This can be seen in Munich and Stockholm, where efficient control of urban sprawl has led to an increase in population density and the use of public transit and a decrease in the use of automobiles.

c) **Social consequences**

From a social perspective, urban sprawl makes it possible to separate income from residential development; therefore, it can create economic and social divisions in the city.

Given the fact that transportation, density and land use are factors affecting urban sprawl, as they are key indicators in the smart growth approach, the effects of sprawl on these factors in three dimensions (i.e. environmental, economic and social) are summarized in Table 2.3.

TABLE 2.3 The effects of urban sprawl on environmental, economic and social aspects

Dimensions	Consequences of transportation changes	Consequences of density changes	Consequences of conversion of rural lands to urban utilities
Environmental	– Increased energy use for transportation – Increased air pollution – Increased surface runoff pollution – Increased noise pollution – Increased land use and fragmentation and disintegration of ecosystems	– Increased residential energy consumption – Fragmentation and disintegration of ecosystems	– Increased land use and stopping of agricultural production from fertile lands – Fragmentation and disintegration of ecosystems – Reduced perspective of quality on the outskirts of the city
Economic	– Increased cost of preparation and maintenance of infrastructure – Increased cost of production and maintenance and use of vehicles – Increased cost of improving and fixing network access problems	– Reduced access to local services	– Increased cost preparation and maintenance of infrastructure – Increased land prices and increased land use changes and hoarding – Increased land taxes in areas located in scattered urban areas
Social	– Increased vehicle accident – Psychological stresses and increased wasted time for trips	– Increased per capita living space – Reduced social relationships and feeling of being together – Decreased sense of place	– Increased urbanization of rural communities – Increased spatial separation and reduced quality of social conditions within the city

Source: Couch & Leontidou, 2007

Urban sprawl vs. smart growth

A huge amount of research has been conducted to address the negative impacts of urban sprawl, suggesting smart growth as the most appealing alternative. However, some argue that urban sprawl can have some benefits for people that should not be simply ignored (Bhatta, 2010). The personal benefits of sprawl include meeting consumers' interest in more socially and economically separated communities with lower housing densities and possibly less housing costs (Wassmer & Bass, 2004)

In support of sprawl, Glaeser and Kahn (2004) point to cheaper and larger houses as a positive aspect. Burchell (2000) believes that the lands far away from the centre of a metropolitan area are cheaper and, as a result, lead to cheaper housing. In response to critics of scattered land use patterns in the United States, Gordon and Richardson (2000) insist that Americans are demanding larger homes and fields that are most likely to be found in remote areas.

Such views on the disadvantages and advantages of sprawl and smart growth indicate the need for accurate analysis of the actual effects to determine the related problems, as well as the main ways of dealing with them. In order to achieve wide acceptance and implementation of smart growth, an analysis of the positive effects for all stakeholders is necessary. Downs (2005) believes that successful implementation of smart growth requires the adoption of policies that are in contradiction with long-standing traditions, such as living in low-density areas. Therefore, the interpretation of the development perspective is important in helping to identify the negative effects of sprawl on the city's integrated body, as well as the long-term interests of groups affected by development. The implementation of smart growth creates different opportunities and negative side effects for different stakeholders. For example, real estate owners on the outskirts of urban areas are likely to incur losses as a result of smart growth because the value of their properties will decrease with limited development in the suburbs. On the other hand, owners of old properties will benefit from an increase in the value of their properties (Downs, 2005). Effective implementation of smart growth requires that all affected stakeholders be taken into consideration in order to increase the chances of adopting regulations for this type of development. Indeed, feasibility studies on the potential impacts of smart growth policies on the stakeholders before adopting this approach is essential to prevent a low rate of adoption (Batisani & Yarnal, 2009).

Peiser (2001) believes that smart growth is a reaction to sprawl. Sprawl has been criticized for increasing housing costs and traffic congestion and for imposing unnecessary infrastructure costs, whereas the goal of smart growth is to balance the needs of individuals and the benefits of economic development (Peiser, 2001). Burchell (2000) defines smart growth as a development pattern that is in opposition to sprawl and is characterized by a more controlled movement outward to direct part of the growth into the metropolitan areas. This approach consumes less capital and fewer natural resources and helps with the realization of more ambitious development (Burchell, 2000). Ruthe (2006) also suggests that sprawl is associated with less dense land use patterns and increased energy consumption, and is dispersed and automobile dependent. However, smart growth is often accompanied by compact development patterns, access to public transit and land reuse and is human-centred and mixed (Ruthe, 2006).

In general, according to Litman (2005), smart growth refers to the principles of development and planning operations, which makes land use and transport patterns more effective. This approach incorporates countless strategies that result in more accessibility, more efficient land use patterns and a hybrid transport system. Smart growth has been proposed to counter sprawl and has been supported by various groups (Litman, 2005).

At the same time, smart growth plans should encourage diversity in transportation options and reduce automobile dependency. Perhaps the most controversial claim in this regard is that the cost of infrastructure projects that create sprawl must be paid for by developers who are constructing in urban areas following a sprawl pattern, rather than being distributed among all citizens (Gray, 2005). Smart growth is opposed to sprawl, focuses on urban regeneration and the expansion of public transport options and seeks to build communities where people are willing to live. This type of growth, with the goal of building a community with a unique concept of place and emphasis on the minimum use of cars, seeks a high environment perception and legibility.

One of the important pillars of smart growth is access, which is defined as the ability of people to reach the desired goods, services and activities. Whereas sprawl emphasizes the "ability to move" (physical motion) and "car rides" (traveling by personal car), smart growth seeks to reduce the distance between activities (home, work, schools and services) and supports various modes of transportation (walking, biking, buses, etc.). As a result of urban sprawl, daily journeys are made longer and have to be done by personal cars, whereas smart growth makes daily trips shorter and provides a variety of transportation options for citizens. A comparison between the two models shows that smart growth has more benefits than does sprawl or horizontal growth and can better contribute to improving the quality of life in cities. Instead of developing cities outward, smart growth tries to expand cities within existing frameworks so that the expansion of cities is prevented, and this helps improve the quality of the community. For these reasons, urban development trends are moving toward this model, and an attempt is made to follow smart growth principles in the expansion and growth of cities (Litman, 2015). As Litman (2015) writes:

> Some argue that scattered growth involves numerous economic, social and environmental costs and that smart growth is desirable. In contrast, the critics of smart growth argue that scattered growth brings benefits that can offset these costs and meet consumer demands.

Chapter 4 of this volume discusses in detail the responses to the most important criticisms of smart growth. Table 2.4 summarizes comparisons between smart growth and urban sprawl.

TABLE 2.4 Responses of smart growth to urban sprawl

Indicator	Smart growth	Sprawl
Density	Higher density, more compact activities	Lower density, diverse activities
Growth pattern	Inner development (brown field) and development in abandoned lands	Development in the periphery of cities (green field) and development on green lands
Use mixing	Mixed uses	Separate uses
Scale	Human scale; smaller buildings, blocks and roads; paying attention to details because people look at nearby scenery like walkways	Large scale; larger blocks and buildings; wider roads; less detail, because people, especially the riders, look at distant landscapes
Public services (shops, schools and parks)	Local, distributed and smaller; good pedestrian access	Regional, in a single location, larger scale; requires automobile access
Transportation	Multi-modal use and transport patterns that support pedestrians, cyclists and public transport	Automobile-oriented use and transport patterns; poor places for walking, cycling and transit
Connections	More connecting roads and walking paths and more direct trips through automotive and non-automotive methods	Hierarchical road network with many unconnected roads and walkways and obstacles for non-traditional road trips
Street design	Gathers a variety of activities and decreases street traffic	Increase the volume and speed of motor vehicle traffic
Planning process	Planning and coordination among legal authorities and investors	Without planning and with low coordination between legal authorities and investors
Public spaces	Emphasis on public areas (street landscape, walking areas, public parks and public facilities)	Emphasis on private territories (yards, shopping malls, community entrances, private halls)

Source: Litman, 2015

Conclusion

The aim of this chapter was to better understand urban sprawl and the reasons for its opposition to smart growth. One of the most important questions raised in this chapter was about the nature of urban sprawl, and the discussion started with a brief history of how this term emerged in urban planning literature. As noted earlier, urban sprawl in the early stages of its formation referred to the physical (horizontal) growth of the city into undeveloped suburbs. The phenomenon of urban sprawl is the offspring of the Industrial Revolution and its associated technologies and was not possible before due to transportation restrictions (pedestrians and chariots). Nevertheless, further studies showed that the development of technology and the invention of motor vehicles alone could not result in urban sprawl, but rather the increase in population, congestion of urban centres and personal preferences diminished the attractiveness of urban centres and increased the desirability of living in the suburbs and open fields.

This study then discussed the causes of sprawl to determine what factors contributed to its development. Investigations showed that the growth of the metropolitan population, the abundance of land, decentralization of employment in urban centres, the ideal housing image, the development of transport and public policies are among the main causes for the prevalence of urban sprawl.

In order to better understand the phenomenon of urban sprawl, this chapter investigated its various aspects and criticized this developmental pattern in terms of economy, density, land use patterns and morphological patterns. It also addressed the implications of urban sprawl in terms of impact location (on cities and suburbs) and also the type of impact (i.e. environmental, economic and social).

In the final section, the contrast between the two approaches of urban sprawl and smart growth was discussed. As mentioned, the smart growth approach is a decisive response to urban sprawl and is struggling to turn the automobile-oriented city into a human-oriented one. In the end, for a better understanding of this chapter, the goals and policies of urban sprawl versus smart growth are presented in Table 2.5.

TABLE 2.5 Review of urban sprawl vs. smart growth policies

Urban sprawl	Smart growth
Economically mass production	Production of economic, cultural and social value
Historical detachment	Restoration of urban history
Rapid growth	Gradual growth
Design of building masses	Design of urban spaces
Movement and speed	Pauses to experience the urban space
Riding city	Walking city
City dominates nature	City coordinated with nature
Design of residential complexes	Design of vibrant neighbourhoods
Design based on bird's eye view	Design is based on pedestrian view
Decentralized and non-congested	Compact with spatial harmony
Maximization of speed and the capacity of roads and streets for personal vehicles	Focus on other modes of transport and the development of public and hybrid transport
Abundant free or cheap parking	Reduced parking space alongside improved parking management
Mega land blocks and increased role of motor vehicles	Small blocks with emphasis on pedestrian access

(Continued)

TABLE 2.5 (Continued)

Urban sprawl	Smart growth
Single-use zoning	Mixed-use zoning
Development on green land; development policies directed at open spaces	Development of brown field, abandoned and worn-out areas; development policies directed at inner parts of the city
Zoning includes large-scale land lots	Division of land into various and small-scale lots
Separation of various types of transport with low connectivity	High diversity and connection in a variety of transport modes
High public transit fares	Affordable public transit fares
Low density in terms of use	High density in allocation of land to uses
Poor walking and cycling conditions	Improved walking and cycling conditions

Note

1 The 2010 UN-HABITAT report, "2010/2011 State of the World's Cities Report, Bridging the Urban Divide", examines the growth of cities and its relation to population growth. For example, the area of Guadalajara in Mexico had a horizontal growth of 1.5 times its population during the 1970s. This was generalized to Johannesburg, South Africa's largest economic pillar; Antananarivo, Madagascar's capital; and Cairo and Mexico City, the capitals of Egypt and Mexico, respectively.

References

Banai, R., & DePriest, T. (2014). Urban sprawl: Definitions, data, methods of measurement, and environmental consequences. *Journal of Sustainability Education, 7.*

Batisani, N., & Yarnal, B. (2009). Uncertainty awareness in urban sprawl simulations: Lessons from a small US metropolitan region. *Land Use Policy, 26*(2).

Baum, H. (2004). Smart growth and school reform: What if we talked about race and took community seriously. *Journal of the American Planning Association, 70*(1).

Bengston, D. N., Fletcher, J., & Nelson, K. (2004). Public policies for managing urban growth and protecting open space: Policy instruments and lessons learned in the United States. *Landscape and Urban Planning, 69.*

Bhatta, B. (2010). *Analysis of urban growth and sprawl from remote sensing data.* New York: Springer.

Black, A. (1991). The Chicago area transportation study: A case study in rational planning. *Journal of Planning and Environmental Review, 10*(1).

Burchell, R. (2000). *Costs of sprawl 2000.* Washington, DC: National Academy Press.

Carruthers, J. I., & Ulfarsson, G. F. (2002). Urban sprawl and the cost of public services. *Environment and Planning B.*

Chen, H. (2008). Sustainable urban form for Chinese compact cities: Challenges of a rapid urbanized economy. *Habitat International, 32*(1).

Chin, N. (2002). *Unearthing the Roots of Urban Sprawl: A Critical Analysis of Form, Function and Methodology.* Retrieved from London.

Clawson, M. (1962). Urban sprawl and speculation in suburban land. *Land Economics, 38*(2).

Clawson, M., & Hall, P. (1973). *Planning and urban growth: An Anglo-American comparison.* Baltimore: Johns Hopkins.

Couch, C., & Leontidou, L. (2007). *Urban sprawl in Europe*. Oxford: Blackwell Publishing.

Downs, A. (2005). Smart growth: Why we discuss it more than we do it. *Journal of the American Planning Association, 74*(4), 367–380.

Duany, A. (2002). Introduction to the special issue: The transect. *Journal of Urban Design, 18.*

Environmental Protection Agency. (2013). *Strategies for advancing smart growth, environmental justice, and equitable development*. Retrieved from USA: https://www.archdaily.com/343439/u-s-epa-strategies-for-advancing-smart-growth-environmental-justice-and-equitable-development

European Environment Agency. (2006). Urban sprawl in Europe. Retrieved from www.eea.europa.eu/publications/urban-sprawl-in-europe

Ewing, R. (1997). Is Los Angeles-style sprawl desirable? *Journal of the American Planning Association, 63*(1).

Ewing, R., Hamidi, S., & Grace, J. B. (2016). Urban sprawl as a risk factor in motor vehicle crashes. *Urban Studies, 53*(2).

Ewing, R., Kostyack, J., Chen, D., Stein, B., & Ernst, M. (2013). *Endangered by sprawl*. Washington, DC: National Wildlife Federation.

Ewing, R., Pendall, R., & Chen, D. (2002). *Measuring sprawl and its impacts*. Washington, DC: Smart Growth America.

Gillham, O. (2002). *The limitless city: A primer on the urban sprawl debate*. London: Island Press.

Glaeser, E., & Kahn, M. (2004). Sprawl and urban growth. *Regional and Urban Economics, 56*(4).

Google Map. (2018). Las Vegas city image. Google Map.

Gordon, P., & Richardson, H. (1997). Are compact cities a desirable planning goal? *Journal of the American Planning Association, 17*(5).

Gordon, P., & Richardson, H. (2000). *Critiquing sprawl's critics*. Retrieved from Washington: https://object.cato.org/pubs/pas/pa365.pdf

Gray, R. (2005). *Growing expectations: Understanding the politics of smart growth in the American states*. College Park, MD: University of Maryland.

Hadley, C. (2000). Urban sprawl indicators, causes and solution. *Urban Studies, 11.*

Halleux, J.-M., Marcinczak, S., & Krabben, E. v. d. (2012). The adaptive efficiency of land use planning measured by the control of urban sprawl. The cases of the Netherlands, Belgium and Poland. *Land Use Policy, 29*(16).

Harvey, R., & Clark, W. A. V. (1965). The nature and economics of urban sprawl. *Land Economics, 41*(1), 1–9.

H.C. Planning Consultants. (1999). *The costs of suburban sprawl and urban decay in Rhode Island*. Rhode: Grow Smart Rhode Island.

Hess, G. R. (2001). Just what is sprawl, anyway? *Carolina Planning, 19*(1).

Hughes, J. W., & Seneca, J. J. (2004). *The beginning of the end of sprawl?* New Brunswick: Rutgers University.

Ibata, D. (2000). *Builders warm to 'smart growth.' Chicago tribune*. Chicago: Real Estate.

Jackson, K. T. (1985). *Crabgrass frontier: The suburbanization of the United States*. New York: Oxford University Press.

Jacobs, J. (1961). *The death and life of great American cities*. New York: Random House.

Jensen, D. (2005). Cincinnati suburb: Wikimedia.org.

Lewyn, M. (2000). Suburban sprawl: Not just an environmental issue. *Marquette Law Review, 84*(2).

Litman, T. (2005). *Evaluating criticism of smart growth*. Victoria, CA: Victoria Transport Policy Institute.

Ludlow, D. (2006). *Urban sprawl in Europe: The ignored challenge*. Copenhagen: J.R.C & European Environmental Agency.

Malizia, E., & Exline, S. (2000). *Consumer preferences for residential development alternatives.* Retrieved from Chapel Hill.

McCann, B. A., & Ewing, R. (2003). *Measuring the health effects of sprawl: A national analysis of physical activity, obesity and chronic disease.* Retrieved from USA: http://smartgrowth.umd.edu/assets/ewingmccann_2003.pdf

Menon, N. (2004). Urban sprawl, vision? *The Journal of the WSC-SD, 2*(3).

Mills, M. E. (2003). Book review of urban sprawl causes, consequences and policy responses. *Regional Science and Urban Economics, 33.*

Mokhtari, R., Hosseinzadeh, R., & Alizadeh, A. S. (2013). Analysis of urban smart growth patterns in fourteen regions of Isfahan based on regional planning models. *Journal of Urban and Regional Studies, 19.*

Newman, P. W. G., & Kenworthy, J. R. (1999). *Sustainability and cities: Overcoming automobile dependence.* Washington, DC: Island Press.

Norman, J., MacLean, H. L., & Kennedy, C. A. (2006). Comparing high and low residential density: Life-cycle analysis of energy use and greenhouse gas emissions. *Journal of Urban Planning and development, 132*(1).

Nozzi, D. (2003). *Road to ruin: An introduction to sprawl and how to cure it.* Westport: Greenwood Publishing Group.

Office of Technology Assessment. (1995). *The technological reshaping of metropolitan America.* Washington, DC: Congress of the United States.

Ottensmann, J. R. (1977). Urban sprawl, land values and the density of development. *Land Economics, 53*(4).

Parisi, T. (1998). Urban sprawl harvesting prime farmland. *State Journal-Register, B5.*

Peiser, R. (2001). Decomposing urban sprawl. *Town Planning Review, 72*(3).

Poelmans, L., & Rompaey, A. V. (2009). Detecting and modelling spatial patterns of urban sprawl in the Flanders-Brussels region. *Landscape and Urban Planning, 93.*

Robinson, L., Newell, J. P., & Marzluff, J. M. (2005). Twenty-five years of sprawl in the Seattle region: Growth management responses and implications for conservation. *Landscape and Urban Planning, 71.*

Ruthe, M. (2006). *Smart growth and climate change: Regional development, infrastructure and adaptation.* Cheltenham: Edward Elgar Publishing Limited.

Seifuddini, F., & Shorjah, M. (2014). *Smart planning of land use and urban transportation, a dialectical look at urban space.* Tehran: Modiran −e- Emrooz Press.

Shankbone, D. (2008). Suburbia: Wikimedia.org.

Siembab, W. (2005). *Retrofitting sprawl: A cyber strategy for livable communities* (01–16). Retrieved from.

Sierra club. (2001). *Population growth and suburban sprawl.* Washington, DC: Sierra club.

Simon, P. (2009). Sprawl in Milton, Ontario. Canada: Wikipedia.org.

Smart Growth Network. (2006). *This is smart growth.* Princeton: Princeton Junction.

Snyder, K., & Bird, L. (1998). *Paying the costs of sprawl: Using fair-share costing to control sprawl.* Washington, DC: The U.S. Department of Energy.

Speir, C., & Stephenson, K. (2002). Does sprawl cost us all? Isolating the effects of housing patterns on public water and sewer costs. *Journal of the American Planning Association, 68.*

Squires, G. D. (2002). *Urban sprawl: Causes, consequences, & policy responses.* Washington, DC: The Urban Institute Press.

Stephens, R., & Wikstrom, N. (2000). *Metropolitan government and governance: Theoretical perspectives, empirical analysis, and the future.* New York and Oxford: Oxford University Press.

Tabibian, M., & Asadi, I. (2009). Investigating and analyzing the dispersed factors in the development of space in metropolitan areas. *Naameh Memari and Shahrsazi Journal, 1*(1), 5–24. doi:10.30480/aup.2009.213

The Real Estate Research Corporation. (1974). *The costs of sprawl.* Washington, DC: U.S. Government Printing Office.

U.S. Department of Agriculture, E. R. S. (2001). *Smart growth: Implications for agriculture in urban fringe areas.* Retrieved from http://www.ers.usda.gov/publications/AgOutlook/April2001/ao280.pdf

U.S. Department of Housing and Urban Development. (2003). *Smart growth and livable communities resources.* Retrieved from http://www.hud.gov/offices/cpd/economicdevelopment/programs/rc/resource/smartliv.cfm

Uhel, R. (2006). Urban sprawl in Europe: The ignored challenge. *EAA Report, 10*(2), 56.

United Stated Department of Agriculture. (2000). *1990–2000 Decennial census of population and housing.* Retrieved from Washington.

Urban Sprawl Inc. (2015). *Urban sprawl in America.* Retrieved from https://www.linkedin.com/topic/urban-sprawl

USHUD. (1999). *The state of the cities 1999: Third annual report.* Retrieved from Washington, DC.

Vermont Forum on Sprawl. (1999). *Sprawl defined.* Retrieved from http://www.vtsprawl.org/sprawldef.htm

Wang, F. L. (2004). Reformed migration control and new targeted people. *The China Quarterly, 177.*

Wassmer, R. W. (2002). *An economic perspective on urban sprawl: With an application to the American west and a test of the efficacy of urban growth boundaries.* Sacramento: California State University.

Wassmer, R. W., & Bass, M. C. (2004). *Sprawl's impact on urban housing prices in the United State.* Sacramento: California Senate Office of Research.

Whyte, W. H. (1993). *The exploding metropolis.* Berkeley: University of California Press.

3

A REVIEW OF GLOBAL EXPERIENCES IN EVALUATING URBAN DEVELOPMENT PLANS AND POLICIES BASED ON SMART GROWTH

Introduction

In this chapter, the practical experiences of smart growth–based urban planning and design first in the United States – as the origin of this movement – in six different cases and then two examples from China, India and Iran are discussed, and the correspondence of the principles, criteria or objectives of each case with the 11 principles of smart growth will be investigated. This chapter covers experiences in Asia, the Middle East and America, and the cases were selected based on accessibility to data and relation to smart growth principles.

North America

The Idaho experiences

Idaho, in the United States, is one of the states where planning is based on smart growth. Examples of this kind of planning in Idaho include the following (Clegg, 2008):

- Ada Urban Areas Highway District emphasizing the provision of various transportation options;
- Special plan of the Harris Ranch community in Boise and Ten Mile in Meridian with an emphasis on mixed land uses;
- The Garden City plan focusing on constructing affordable housing;
- The Sandpoint plan emphasizing the creation of pedestrian-oriented neighbourhoods;
- The plans for cities of Caldwell, McCall and Greenleaf, with an emphasis on producing a sense of place with social attractiveness.

Given the importance and thematic relevance to the content of this book, the case studies related to the two principles of "providing a variety of transportation choices" and "mixing land uses" in the state of Idaho are further discussed next.

Ada Urban Areas Highway District emphasizing the provision of various transportation choices

The first principle mentioned for smart growth planning in this state is that of "providing various transportation choices". As reported, the excessive use of personal vehicles in all types of trips has given rise to a "multiplicity of transportation modes". Previous studies of this report show that 79 percent of trips to destinations such as work, shops, leisure or education are carried out with a personal vehicle. Also, according to the studies conducted, due to the lack of suitable sidewalks, even short distances between public spaces are not desirable enough to attract pedestrians, leading to an increased use of cars. Providing safe and reliable public transport, proper sidewalks, continuous and coherent bicycle and pedestrian lanes, etc. can increase health, reduce energy consumption and protect the environment (Clegg, 2008).

The plan developed on a ten-year horizon for the Ada urban area in Boise in 2005 emphasized bicycle- and pedestrian-oriented transportation. The plan focuses on the design and redevelopment of the existing and future bicycle lanes and roads. The total cost of the plan is estimated at $362.5 million. The most important point in this plan is prioritizing the repair of existing pathways and creating new ones.

The most important criteria to consider when designing new paths are:

- Filling the empty intervals between available bicycle and pedestrian paths;
- Placing paths at a radius of approximately 402 metres from schools, public places and parks;
- Creating a bicycle path along the existing river valleys;
- Providing proper access to shopping malls;
- Creating bicycle paths alongside first-grade arteries.

Each of these criteria has played a key role in prioritizations for upgrading and building new paths (The Transpo Group, 2005).

Harris Ranch plan in Boise and Ten Mile plan in the city of Meridian with an emphasis on mixing land uses

The second principle in the smart growth urban plan for Idaho is "mixing land uses". In this plan, the emphasis has been placed on mixing land uses in the construction of pedestrian-oriented neighbourhoods.

By building stores, offices and residences to be close to (or on top of) each other in appropriate locations, the City allows people to work, shop and enjoy recreation close to where they live. It makes driving trips shorter, transit, walking and biking more convenient (encouraging a healthier lifestyle), protects the environment, lowers transportation costs and conserves energy (Idaho Smart Growth, 2014).

The cities in the state of Idaho, where special attention has been paid to planning with an emphasis on mixing land uses, include (Clegg, 2008):

- *The city of Boise – Harris Ranch plan (2006): adoption of special plans for urban areas and planning for mixed-use areas.*

In 2006, a plan was developed for the Harris Ranch area of Boise, in which a special emphasis was placed on mixing urban land uses. Harris Ranch will achieve a true mixed-use community using the Specific Plan Ordinance adopted by the City of Boise. The plan calls for relative high density, good jobs and housing balance, diversity in housing and a well-connected transportation system that encourages walking, biking and transit while offering efficiency for drivers. Housing densities range from 3 units/acre at the edges to 12 units/acre in the core. Development is centred around a new "main street" along Warm Spring Avenue with a mix of commercial, retail and urban style housing. The Specific Plan Ordinance ensures this balance through carefully crafted and unique zoning requirements that were adopted as part of the specific plan. Those requirements include defined percentages of residential to commercial development in each phase of the overall plan and detailed form-based design standards. As Boise's first Specific Plan, Harris Ranch chartered new waters for development in the city (Boise City Council, 2006).

- *The city of Meridian – Ten Mile Plan (2007): adopting special (topical-thematic) plans for districts and planning mixed-use areas.*

This plan will guide the future development of the Ten Mile Interchange area in a manner that is acceptable and beneficial to all interested stakeholders. The Plan focuses on matters of land use, access, mobility, sustainability and quality and emphasizes transit-oriented development patterns. The Ten Mile Interchange Area will look, feel and function differently than a typical commercial area or residential subdivision. Many residential uses will occupy the second and third levels of buildings above retail, office and light industrial uses on the ground floor. In some commercial areas, residential uses may occupy the lowest levels of buildings. Unlike many commercial and employment districts, the Ten Mile Interchange Area will not empty out at 5 p.m. when employees leave work. For many employees, home will be upstairs, around the corner or down the street. This area allows a range of land uses – from industrial to residential to commercial – in close proximity to one another. This mix, anchored by a lifestyle centre, will create an exciting atmosphere for residents and a unique new area of Meridian (Colliers International, 2007).

The California experience

Riverside is located in the growing metropolitan area of "Inland Empire" in the south of California. The plan for this area was developed in 2003 by the Western

Riverside Council of Governments (WRCOG). The degree of viability of smart growth has been evaluated in this plan (Hogle-Ireland, 2007).

Large-scale construction in the southern parts of the California coast in Los Angeles and Orange led to progress and yet destruction of the natural environment around the city. This progress continued to the extent that only scattered parts from the basin of the city of Los Angeles remained (City of Los Angeles, 2014). Following these problems, most people, especially young families, were driven out of the area due to high housing costs, while many of them worked in coastal areas. The main reason for this relocation was to access affordable housing. Based on a plan prepared for Western Riverside, in order to evaluate the smart growth–based development, the ten smart growth principles were re-classified into four main categories (Hogle-Ireland, 2007):

- Land use;
- Housing and employment;
- Public transportation;
- Infrastructure.

A number of sub-criteria are defined for each of these four principles, allowing for their quantitative evaluation and analysis (Table 3.1).

TABLE 3.1 Assessment of urban development based on the four smart growth principles in the Riverside District of California

Principles	Sub-criteria	Description
Land use	Compact development	Compact residential development provides more open spaces, reduces the number of trips by cars, reduces the costs of infrastructure, results in services being deployed at a shorter distance from residences and gives citizens the right to have more choices in the housing market, leading to the settlement of more people.
	Providing the minimum amount of land needed in single-family residential areas	There are problems with large-scale land lots, including: – More land is allocated to the construction of each residential unit; – The distances of intra-city trips increase; – Houses are usually placed far from services; – A smaller number of people are located in the area; – These areas are costly in terms of development.
	The development of mixed-use areas	Deploying different and adaptable uses in close proximity to each other allows people to simultaneously live, work, shop and engage in leisure activities in the same area.

(Continued)

TABLE 3.1 (Continued)

Principles	Sub-criteria	Description
	Stressing infill development	Infill development refers to the construction of houses, commercial centres and public facilities on the unused land within the city. In this type of development, the existing infrastructure is used, and there is no need to spend on developing new infrastructure. The goals of infill development are aligned with the goals of increasing density and mixing land uses. Maximum use of land within urban areas reduces the need to develop "empty and undeveloped land" in the new areas and to expand the city outward.
	Construction of parks (per 1,000 inhabitants)	Parks improve the quality of life in any city, and they will be most efficient when they are not too crowded and are accessible on foot. Hence, it is better to build a parkland per 1,000 people in neighbourhoods.
Jobs and housing	Affordable housing	Affordable housing refers to housing that can be purchased by no more than 30 percent of the gross household income. This requires the existence of different housing choices at reasonable prices for different income levels.
	Jobs–housing balance	Jobs–housing balance is the ratio of the number of job opportunities within a defined area to the number of residential units occupying the same area. This indicator shows the amount of people traveling outside the neighbourhood to go to work. If this ratio is negative, it will result in: – More intra-city travels; – More economic emphasis on the services sector; – Low tax base; – Poor investment in local schools; – Increasing need for affordable housing.
	Accessory dwelling units (ADUs)	ADUs are independent residential units where residents benefit from a shared space. The use of such units can increase the density.
Public transport	Desirable public transport	This indicator measures the level of comfort and security of people in using public transport systems. If public transport is slow, unordered, insecure, inaccessible and unpleasant, people will not use it. The following are features of popular and desirable public transport: – Frequent and regular services; – Safe and convenient stations; – Appropriate destinations; – Passengers' desire to use the public transport system; – Reasonable fares.

Principles	Sub-criteria	Description
	Compliance with parking standards in residential and commercial (retail) areas	Allocating larger spaces to parking lots can reduce the amount of land that can be used to create income-generating spaces, upgrade walking paths, create a desirable open-space landscape plan or create houses close together. It will: – Lead to spoiling resources and spaces; – Usually create uninhabited islands on the streets; – Lead to the emergence of places for waste collection; – Cause warming and air pollution in the city.
	Vehicle mile travelled (VMT)	There is a logical relationship between land use and public transport. Theoretically, denser developments require a smaller public transport infrastructure because in this situation, some trips are made without the use of cars. In addition, reducing car travels increases air quality, reduces traffic and increases investment in the public transportation infrastructure.
Infrastructure	Traffic-calming strategies	Traffic-calming strategies make walking easier for pedestrians, slow down the movement of vehicles in local traffic and: – Make walking a safe option; – Prevent unnecessary traffic; – Reduce air and noise pollution; – Promote pedestrian-oriented environments; – Lower the cost of street construction.
	Street network patterns	Different street networks patterns in the design of neighbourhoods significantly affect travelling modes, as well as the dynamics of the neighbourhoods. The popular radial pattern in the suburbs will increase the level of dependence on cars, while the gridiron pattern pays more attention to pedestrians, encouraging people to walk and use bicycles, especially for short trips. There are some differences between the various networking patterns: – Curved networks pay more attention to vehicle traffic; – Gridiron patterns stress pedestrians; – Gridiron patterns place services closer together; – Gridiron patterns reduce noise pollution; – Gridiron patterns are more suitable for a smart growth plan; – Gridiron patterns improve air quality in the long run.

Source: Adapted from: (Hogle-Ireland, 2007)

As mentioned, four general principles are used in the Western Riverside case, but they cannot cover all dimensions of smart growth. In this plan, principles such as the participation of local communities in urban development projects and the adoption of fair and enforceable decisions for the development and creation of distinctive and attractive neighbourhoods with a strong sense of place have been ignored. A point to consider in this case is the statement of general principles, each with different sub-criteria. For example, regarding mixed uses, infill development and traffic calming, instead of defining more precise and objective indicators, indicators such as rules, incentives and policies are considered in each case. One of the reasons for this is that Western Riverside is composed of 14 cities, each with its own specific characteristics, and it is not possible to evaluate smart growth in those cities with more precise indicators. In other words, each city may achieve positive results in smart growth with different policies. Smart growth does not specifically define a strict framework so that if a community falls within the framework, we can claim that it follows and benefits from smart growth, and if it fails to do so, it falls short of achieving smart growth objectives. Indeed, in the smart growth approach, it is possible to determine various strategies and programs for implementation of the ten principles and give more decision-making freedom to communities and authorities. Table 3.2 categorizes the goals set out in the Riverside plan based on the ten principles of smart growth.

TABLE 3.2 The classification of the criteria considered in the Riverside plan based on the ten smart growth principles

Criteria (based on the ten principles of smart growth)	Sub-criteria	Smart growth measure and process
Principle 1. Mixing land uses	Diversification and integration of uses	The city can consider certain incentives for mixing uses, including: – Allowing surplus density; – Speeding up processes; – Reducing permits and visiting fees; – Collaborating in creating the required infrastructure.
Principle 2. Compact development	Provision of the minimum amount of required land	In villa areas, up to an approximate area of 372 sq m or less with the possibility of including 6 units more than 11 residential units in each 0.4 hectare.
Principle 3. The provision of various housing choices	Housing affordability	Increasing the number of families that can afford to pay rentals.
	Ratio of jobs per household	Increasing this ratio to more than 1.5.
	ADUs	Licensing of residence in these units by the city authorities.

Criteria (based on the ten principles of smart growth)	Sub-criteria	Smart growth measure and process
Principle 4. Creating walkable neighbourhoods	Traffic calming	The city must have policies, laws and incentives to expand traffic-calming measures that are regularly applied in the community.
	Narrow streets	The pathways should have a minimum width of 8.5 metres.
Principle 5. Fostering distinctive, attractive communities with a strong sense of place	–	–
Principle 6. Preserving open space, farmland, natural beauty and critical environmental areas	Creating and developing parks and recreation green spaces	Constructing at least one park per 1,000 inhabitants up to more than five hectares.
Principle 7. Strengthening and directing development towards existing communities (infill development)	Paying attention to infill development policies and incentives	The cities should provide specific incentives to support infill development, including: – Permitting surplus density; – Speeding up processes; – Reducing permits and visiting costs; – Collaborating in providing the required infrastructure.
Principle 8. Providing a variety of transportation choices	Street networking patterns	Using a gridiron street network pattern in street design:
	Popular and desirable transport; passengers' access to public transport	– Locating the minimum number of bus stops on the route; – More than 10 percent of the bus stops should have sunshades. – More than 25 percent of bus stops should have seats.
	Paying attention to the driving distances in the city	Reducing the travel distance by reducing the maximum length of intra-city driving distances.
	Compliance with parking requirements and standards in apartment and retail areas	Fewer than 2 parking units are required per 93 square meters of commercial space and fewer than 1.25 parking units per two-room residence.
Principle 9. Making development decisions predictable, fair and cost-effective	–	–

(Continued)

TABLE 3.2 (Continued)

Criteria (based on the ten principles of smart growth)	Sub-criteria	Smart growth measure and process
Principle 10. Encouraging community and stakeholder collaboration in development decisions	–	–

The Maryland experience

Since 1996, the Development, Community and Environment Division (DCED) has begun to work with national organizations, local governments, universities and the private sector to support smart growth policies. It was concluded during these collaborations that the assessment of smart growth requires being able to measure the natural environment status, public transport, and quality of life over a short period. To this end, United States Environmental Protection Agency (EPA) introduced the Smart Growth Index (SGI). The SGI is a software tool that allows the user to assess the existing environment and community conditions, compare the effects of multi-step development and different public transport scenarios and control changes over time. It has made the comparison of these effects simple through the use of transparent map sheets. The original version of the SGI was prepared in 2000. The SGI model is a fast and multi-faceted analytical tool in response to "What if?" scenarios for communities. SGI is based on a geographic information system (GIS), and its output includes maps, charts and tables. SGI helps communities evaluate different development scenarios by rating projects through the following measurable indicators (Office of Smart Growth, 2005):

- Population density (persons per square meter);
- Land use mix;
- Land use diversity;
- Housing density;
- Single-family housing share (percentage of single-family units/total dwellings);
- Multi-family housing share (percentage of multi-family units/total dwellings);
- Housing–transit proximity (percentage of dwellings within 2.25 km of transit stops);
- Housing–recreation proximity (percentage of dwellings within 2.25 km of parks);
- Residential unit energy consumption;
- Residential unit water consumption (cubic meter/day/capita);
- Jobs/housed workers balance (ratio of total jobs to total housed workers);
- Employment density (employees/employment land);
- Employment proximity to transit (percentage of employees within 2.5 km of transit stops);

- Park space availability (park area/1,000 persons);
- Open space area;
- Sidewalk completeness (percentage of streets with sidewalks);
- Pedestrian route directness (average ratio of walking distance from the point of origin to the central node compared to a straight-line distance);
- Street network density (street length [km] separated by neighbourhoods [sq m]);
- Street network connectivity (ratio of intersections to total intersections plus cul-de-sacs);
- Pedestrian environment design;
- Vehicle trips (vehicle trips/day/capita);
- Vehicle distance travelled (distance driven/day/capita);
- Vehicle travel cost (dollars/year/capita);
- Carbon monoxide vehicle emissions (dollars/year/capita);
- Hydrocarbon vehicle emissions (dollars/year/capita);
- Sulphur oxide vehicle emissions (dollars/year/capita);
- Particulate matter vehicle emissions (dollars/year/capita);
- Carbon dioxide vehicle emissions (dollars/year/capita);
- Nitrogen oxide vehicle emissions (dollars/year/capita).

The Maryland Department of Planning (MDP) was one of 20 selected organizations using the SGI. The model was used in the department's studies for redevelopment of the Baltimore Inner Harbor, also known as the Digital Harbor. Among the Digital Harbor projects, MDP officials decided to focus on the Fells Point project, which included the development of residential, office, commercial, retail, recreation and catering areas. Maryland's city planners estimated that these projects could create 14,800 new jobs and 1,100 new residential units. Using the SGI model, planners were able to estimate the air quality effects caused by the mixing of residential and employment sectors in Fells Point. Hence, some of the environmental performance indicators of this model, which were specifically related to the types of trips and air pollution, were selected. Using the smart growth guide, the planning department estimated that smart growth development in the Fells Point area could reduce the VMT and reduce nitrogen oxide production by 14 percent annually, as well as volatile organic compounds and greenhouse gas emissions, through a construction reduction plan in the area. Table 3.3 lists the indicators considered for this project.

In this case, from among the 29 indicators of the SGI model, those responsive to the objectives of the project in assessing the performance of the Fells Point design were selected. The purpose of this project was to estimate the effects of air pollution caused by the mixing of residential and office uses in the area. Therefore, nine indicators were selected, and the project was compared with the ideal scenario. The project was on a small scale, and therefore more objective and specific indicators were needed for its evaluation. In such circumstances, the indicators also affected the evaluation method. Hence, in this project, evaluation was done relative to an ideal scenario. In Table 3.4, the evaluation indicators in the Maryland State Plan are classified according to the ten principles of smart growth (Office of Smart Growth, 2005).

TABLE 3.3 Sample SGI Indicators in the Fells Point pilot project (Baltimore, Maryland)

Indicator	Description	Ideal scenario[1]	Planned scenario for Fells Point
Persons/sq mi	Persons (residents and employees) per sq mi	100,000	75,570
Jobs/housing balance	Ratio of total jobs to total housed workers	1.0	6.93
Land use mix	Proportion of dissimilar land uses among a grid of one-acre cells	1.0	0.63
Street network density	Length of street in miles divided by areas of neighbourhood in square miles (miles per sq mile)	10	34.6
Sidewalk completeness	Percentage of street frontage with sidewalks	100	100
Route directness	Ratio of shortest walking distance from outlying nodes to neighbourhood centre vs. straight-line distance	1.3	0.9
Street connectivity	Ratio of intersections vs. intersections and cul-de-sacs	1.0	0.67
Average distance to transit stop	Average distance from dwellings to closest transit stop in feet	600 (maximum ideal distance)	229
Housing near transit	Percentage of dwellings within 1/4 mile of transit stops	100	90

Source: Office of Smart Growth, 2005

TABLE 3.4 The classification of indicators considered in the Maryland State project based on the ten principles of smart growth

Indicator (based on the ten principles of smart growth)	Sub-indicator	Smart growth process
Principle 1. Mixing land uses	The ratio of dissimilar uses in a network of one-hectare units	Increasing the ratio of dissimilar uses in a network of one-hectare units up to 1 (ideal scenario)
Principle 2. Compact development	Population density (inhabitants and employees) per square mile	Increasing this figure up to 100,000 persons per sq mile (ideal scenario)
Principle 3. The provision of various housing choices	Ratio of total jobs to total occupied houses in the area	Increasing this figure up to 1 (ideal scenario)

Indicator (based on the ten principles of smart growth)	Sub-indicator	Smart growth process
Principle 4. Creating walkable neighbourhoods	Sidewalk completeness	Increasing streets with sidewalks up to 100 percent (ideal scenario)
	Average ratio of walking distance from the point of origin to central node, compared to straight-line distance	Decreasing this figure up to 1.3 (ideal scenario)
	Housing distance from transit stops	Maximum of 600 meters between dwellings and transit stops
Principle 5. Fostering distinctive, attractive communities with a strong sense of place	–	–
Principle 6. Preserving open space, farmland, natural beauty and critical environmental areas	–	–
Principle 7. Strengthening and directing development towards existing communities (infill development)	–	–
Principle 8. Providing a variety of transportation choices	Ratio of intersections to total intersections plus cul-de-sacs	Increasing this figure up to 1 (ideal scenario)
	Street length separated by neighbourhoods (mile per square mile)	Decreasing this figure up to 10 (ideal scenario)
Principle 9. Making development decisions predictable, fair and cost-effective	–	–
Principle 10. Encouraging community and stakeholder collaboration in development decisions	–	–

The experience of Ontario's Greater Golden Horseshoe smart growth plan

Canadian cities are experiencing rapid population growth and development that have increased the footprint of the country's largest urban regions. A predominance of this growth is taking place on the urban periphery of many of these cities, or is "leapfrogging" away from denser developments, replacing what was previously natural or agricultural land with low-density, single-use developments, or "sprawl" (Blais, 2011; Ruth, 2006). This pattern of growth and development accelerates the loss of natural areas and farmland, increases resource consumption and necessitates infrastructure and servicing improvement, as well as expansion, all at a great cost (Burchell, Downs, McCann, & Mukherji, 2005; Tomalty & Alexander, 2005). If unchecked, this growth will perpetuate sprawl and threaten the economic, environmental and social sustainability of Canadian cities.

The effects of growth are particularly evident within Ontario's Greater Golden Horseshoe (GGH). Projected to have 11.5 million inhabitants by 2031, the GGH is Canada's fastest-growing region. Increased growth has made the region's communities more vibrant and diverse; helped maintain a strong economy; and aided in the expansion of community services, arts, culture and recreation facilities. However, growth has adversely affected traffic congestion, the availability of green space and the cost and quality of public infrastructure (Ministry of Infrastructure, 2011).

In an attempt to mitigate the negative effects of this growth, while preserving the positive, the government of Ontario adopted a program of smart growth and introduced a suite of complementary legislative changes to "control" sprawl, build healthy communities, maintain a strong economy and make more efficient use of land and infrastructure in the region (Ministry of Infrastructure, 2011). An extensive greenbelt was also established under the Greenbelt Act and the Ontario Places to Grow Act, both of which became law in 2005 and formed the legislative centrepiece of the 25-year Growth Plan for the Greater Golden Horseshoe, 2006. This planning framework represents the most promising attempt to address sprawl in Canada and realize smart growth.

The Places to Grow Act gives the province the statutory authority to designate any geographic region of Ontario as a growth plan area and mandates the ministries of energy and infrastructure to prepare specific density targets and planning priorities within them. Local planning decisions, including zoning, must conform to the policies in the growth plan; otherwise, the provincial government has the authority to amend municipal decisions. The Greenbelt Act, 2005, authorized the province to designate a greenbelt area and establish the greenbelt plan to protect approximately 1.8 million acres of environmentally sensitive and agricultural land in the GGH from urban development and sprawl. Within this protected area, about 800,000 acres of land are bounded by the areas designated in the Niagara Escarpment Plan and the Oak Ridges Moraine Conservation Plan (Ministry of Municipal Affairs and Housing, 2015).

Inspired by the smart growth movement emerging in the United States during the mid-1990s, the concept was soon embraced in Canada by both governmental and non-governmental organizations (NGOs). Prior to its integration into Ontario's Growth Plan, however, Smart Growth BC, an independent non-profit group, was set up in 1999 to promote compact urban centres, protect resource lands, ensure adequate affordable housing, promote sustainable transportation and maintain environmental integrity (Smart Growth BC, 2008; Tomalty & Alexander, 2005). Shortly thereafter, a wave of smart growth initiatives was implemented by municipalities, regions, businesses and NGOs throughout Canada, and Ontario built upon their success when the former Conservative government launched its Smart Growth Ontario initiative and established a Smart Growth Secretariat in 2001 (Ministry of Municipal Affairs and Housing, 2015).

Though many smart growth principles were in place in Ontario as policies in local plans across the Toronto metropolitan region – such as transit supportiveness, higher residential densities and maintaining urban boundaries – the concept still held considerable appeal as Ontario's preferred method of addressing the region's growth problems (Eidelman, 2010; White, 2007). To carry out this initiative, the government set up panels of citizens and elected politicians representing different interests and asked them to find solutions to growth-related problems (White, 2007). The result was a final report from the Smart Growth Secretariat that was "long on visions and ideals and short on realistic strategies for achieving them" (White, 2007). In addition, many proponents of smart growth believed the government's use of the term was ill-defined, given that it included a resumption of provincial transit spending, brown field redevelopment and environmental protection, while promoting a major new highway building program (Tomalty & Alexander, 2005). In the end, although efforts were made to introduce smart growth, low-density development continued to push outward at the fringes of Toronto's urban area at unprecedented levels, and often into prime agricultural lands (Sewell, 2009).

The Portland experience[2]

The Portland metropolitan area is admired across the nation for its innovative approach to planning. The region's enviable quality of life can be attributed in part to a stubborn belief in the importance of thinking ahead.

One example of this foresight is the 2040 Growth Concept, a long-range plan that reflects the input given by thousands of Oregonians in the 1990s and adopted by the Metro Council in 1995.

Policies in the 2040 Growth Concept encourage:

1   Safe and stable neighbourhoods for families;
2   Compact development that uses land and money efficiently;
3   A healthy economy that generates jobs and business opportunities;

4   Protection of farms, forests, rivers, streams and natural areas;
5   A balanced transportation system to move people and goods;
6   Housing for people of all income levels in every community.

Ten urban design components are identified in the 2040 Growth Concept as the focal points for growth:

> The central city, or downtown Portland (Figure 3.1), serves as the region's business and cultural hub. Within the region, it has the most intensive development of housing and employment, with high-rise development common in the central business district. Downtown Portland will continue to serve as the region's centre for finance, commerce, government, retail, tourism, arts and entertainment.

Town centres provide services to tens of thousands within a two- to three-mile radius. One- to three-story buildings for employment and housing are characteristic. Town centres have a strong sense of community identity and are well served by transit. They include small city centres, such as Lake Oswego, Tualatin, West Linn, Forest Grove and Milwaukie, and large neighbourhood centres, such as Hillsdale, St. Johns, Cedar Mill and Aloha. The main streets are similar to town centres: a

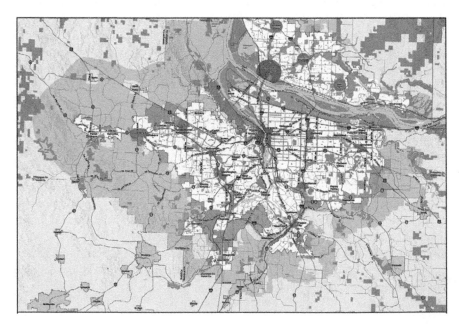

FIGURE 3.1   2040 growth concept map for Portland

Source: Oregon Metro, 2014

traditional commercial identity but on a smaller scale, with a strong sense of the immediate neighbourhood. They feature good access to transit and include:

1   Southeast Hawthorne Blvd. in Portland;
2   Boones Ferry Rd. in Lake Grove;
3   Adair and Baseline streets in Cornelius.

Regional centres are hubs of commerce and local government services serving hundreds of thousands of people. They are characterized by two- to four-story, compact employment and housing development served by high-quality transit. In the plan, eight regional centres become the focus of transit and highway improvements:

•   Gateway, serving central Multnomah County;
•   Downtown Hillsboro and Tanasbourne/AmberGlen, both serving western Washington County;
•   Downtown Beaverton and Washington Square, both serving eastern Washington County;
•   Downtown Oregon City and Clackamas Town Center, both serving Clackamas County;
•   Downtown Gresham, serving eastern Multnomah County.

Station communities are areas of development centred on a light-rail or high-capacity transit station that features a variety of shops and services. These communities are accessible to bicyclists, pedestrians and transit users as well as cars. They include:

1   E 102nd Avenue on the MAX blue line;
2   Cascade Station on the MAX red line;
3   Orenco Station on the MAX blue line.

Neighbourhoods remain largely the same under the 2040 growth concept. Some redevelopment can occur to better use vacant land or under-used buildings. New neighbourhoods are likely to have smaller single-family lots, mixed uses and a mix of housing types such as row houses and accessory dwelling units. The growth concept distinguishes between slightly more compact inner neighbourhoods and the slightly larger lots and fewer street connections of outer neighbourhoods.

Corridors are streets that serve as major transportation routes for people and goods. Extensively served by transit, corridors include:

1   Tualatin Valley Highway and 185th Avenue in Washington County;
2   Powell Boulevard in Portland and Gresham;
3   McLoughlin Boulevard in Portland and Clackamas County.

Industrial areas and freight terminals serve as hubs for regional commerce. They include industrial land and freight facilities for truck, marine, air and rail cargo sites that enable goods to be generated and moved in and out of the region. Access is centred on rail, the freeway system and roadway connections. Keeping these connections strong is vital to a healthy regional economy.

Parks and natural areas are lands that will remain undeveloped, both inside and outside the urban growth boundary. These include parks, stream and trail corridors, wetlands and floodplains.

Rural reserves are large areas outside the urban growth boundary that will remain undeveloped through 2060. These areas are reserved to provide long-term protection for agriculture, forestry or important natural landscape features that limit urban development or help define appropriate natural boundaries for development, including plant, fish and wildlife habitat; steep slopes; and floodplains.

Neighbouring cities are communities such as Sandy, Canby, Newberg and North Plains, which have a significant number of residents who work or shop in the metropolitan area. Cooperation between Metro and these communities is critical to address common transportation and land use issues.

The San Diego Regional Comprehensive Plan[3]

The San Diego Regional Comprehensive Plan (RCP) was developed in 2004 to provide a blueprint for managing the region's growth while preserving natural resources and limiting urban sprawl. The plan is based on the concept that thoughtful land use planning and targeted transportation investments can shape private development to achieve smart growth principles. The San Diego Association of Governments (SANDAG) worked with San Diego area cities and county government to collectively identify urban centres and transit corridors to prioritize for growth. These areas are eligible for grants from SANDAG for planning and infrastructure projects.

The RCP is now also being used as the basis for compliance with the state's recently passed SB 375, a law that requires each metropolitan area to develop a sustainable community strategy to meet a greenhouse gas emission reduction target through land use, transportation and housing planning. SANDAG will be creating the sustainable community strategy as it updates the Regional Transportation Plan (RTP) for 2050.

SANDAG is the metropolitan planning organization and transportation planning agency that represents the 18 cities and the county government in the San Diego region. SANDAG builds consensus; makes strategic plans; obtains and allocates resources; plans, engineers, and builds public transportation; and provides information on a broad range of topics pertinent to the region's quality of life. Although SANDAG does not have regulatory authority over any local governments, its plans and incentives are used by local governments to guide local planning and investment decisions.

The RCP's success is largely due to the active role that the local governments played in its development. SANDAG began by working with the governments to create a consensus definition of smart growth. The association then led the development of a smart growth concept map that officially designates areas that are able to accommodate higher residential or employment density, are pedestrian friendly, and are connected to other centres by transit. Areas can be designated into seven categories based on size and use: metropolitan centres, urban centres, town centres, community centres, rural villages, mixed-use transit corridors and special-use centres.

Cities were asked to identify "existing/planned" areas that either currently meet smart growth principles or are officially slated to be in local master plans. Cities could also identify "potential" areas that hold promise as smart growth centres but are either not currently zoned or planned as such or lack transportation connections. SANDAG used regional growth data to confirm that each nominated area met specific thresholds for density potential before officially designating the area on the smart growth concept map. As a result, the final map represents the priorities of local governments as well as the areas with the greatest potential based on geographic, infrastructure and growth capacity criterion.

In 2004, the same year that the RCP was adopted, county residents approved a 40-year extension of a half-cent sales tax to support smart growth and environmental mitigation projects in the region. This TransNet fund generated $3.3 billion between 1998 and 2008, and the extension is expected to generate more than $14 billion for highway, transit and local road projects. A total 4 percent of the funds (2 percent for each) are set aside for an Environmental Mitigation Program and a Smart Growth Incentive Program.

The TransNet fund generates approximately $4 to 6 million each year for the Smart Growth Incentive Program and is disbursed through a grant process every two years. Local governments can apply for up to $2 million for planning efforts or infrastructure projects. Only designated existing/planned smart growth areas qualify for infrastructure grants, whereas both existing/planned and potential areas qualify for planning grants. Because local governments often leverage matching funds, the grant program is able to fund as many as 20 projects in each cycle. The grants are highly competitive, as they can be used for a wider range of projects and have less strenuous reporting requirements than many state and federal funding programs.

According to the RCP Monitoring Report, early metrics of the results of the RCP include:

- The percentage of new housing units built in Smart Growth Opportunity Areas has increased fitfully, improving from 37 percent in 2004–2005 to 44 percent in 2008–2009.
- From 2005 to 2008, Smart Growth Opportunity Areas experienced a net gain of over 11,000 jobs, while non–Smart Growth Areas experienced a net loss of over 9,000 jobs.

- Regional transit ridership increased steadily from around 89 million riders in 2004 to around 104 million in 2009.
- Travel volumes decreased slightly in most major highway corridors, and commuting travel time decreased on nearly every major roadway between 2005 and 2009.
- The percentage of solid waste that was recycled was close to achieving the state-mandated target.
- Recycled water use continued to increase substantially.

TransNet funds have supported an additional 227 lane-miles of highway, provided financing for more than 800 local road projects and expanded regional transit with 85 miles of trolley and commuter rail lines. The funds have also been used to provide discounted transit passes for disabled people, seniors and students. Selected TransNet projects include:

- San Diego Trolley extended to Santee, Old Town and through Mission Valley to San Diego State University and La Mesa;
- COASTER Commuter Rail opened between Oceanside and San Diego;
- Sprinter light rail opened connecting Oceanside and Escondido.

The Smart Growth Incentive Program has provided around $32 million to 24 capital and six planning projects in its 2005 pilot and its 2009–2010 funding cycle. The next funding round is expected in 2011. Projects funded include:

- Park Boulevard at Harbor Drive Pedestrian Bridge, Centre City Development Corporation;
- Grossmont Trolley Station Pedestrian Enhancements, City of La Mesa and Metropolitan Transit System;
- Mid-City Urban Trail and SR-15 Bikeway, City of San Diego;
- Maple Street Pedestrian Plaza, City of Escondido;
- 8th Street Corridor Smart Growth Revitalization, National City;
- Third Avenue Streetscape Implementation Project, City of Chula Vista;
- Mid-City SR 15 Bus Rapid Transit (BRT) Station Area Planning Diego Study, City of San Diego.

The RCP Monitoring Report does not include information on the reduction of greenhouse gas emissions or energy use by geographic area or by grant allocation area from the TransNet Smart Growth Incentive Program. However, SANDAG is preparing a 2050 RTP built in part off of the RCP, and is evaluating greenhouse gas reductions by transportation scenario for that project. As of December 2010, the draft 2050 Regional Transportation Plan projects that development in accordance with the preferred scenario would reduce per capita carbon dioxide emissions from passenger vehicles by 13 percent from 2005 levels in 2020 and 18 percent by 2035, exceeding the SB 375 targets for the region of 7 percent and 13 percent, respectively. Fuel consumption per day per capita is projected to decrease from 1.45 gallons in 2008 to 0.89 gallons in 2050.

SANDAG has approached the plan's implementation through incentives and collaboration, because it does not have land use authority and is not a regulatory agency. As a result, successful implementation has required collaboration between SANDAG and local government agencies, as well as monetary incentives (like the Smart Growth Incentive Program and Environmental Mitigation Program) and providing resources (like Smart Growth Design Guidelines, Smart Growth Photo Library and Visual Simulations, Smart Growth Trip Generation and Smart Growth Parking Strategies). It has not been a "top-down" effort, but rather a "bottom-up" effort with a focus on collaboration and incentives.

There has been much change since 2004 when the RCP was adopted, including a new focus on climate change (and passage of SB 375 in California), a growing emphasis on public health as an important part of the land use/transportation planning field, a greater emphasis on transit planning as part of the transportation network, stronger relationships with tribal nations in the region and economic cycles that have slowed, but not stopped, development. SANDAG aims to consider these changes in future planning, including the 2050 RTP.

Additionally, SANDAG planners have noticed a considerable increase in receptivity to smart growth and sustainable development principles, with many jurisdictions moving toward higher densities and more mixed uses through general plan and community plan updates of their own accord. The 2050 Regional Growth Forecast shows that most of the new growth to 2050 will take place in the most urbanized areas of the region, with a greater share of multi-family housing growth near transit than ever before.

However, funding remains a concern. For every round of grants that SANDAG issues, many worthy applications cannot be funded. In addition, transit operation funding has been cut from the state budget on numerous occasions. As transit services continue to be cut, it is challenging to talk about smart growth – especially the connection between higher densities and increased transportation choices – when local communities are feeling the cuts and worried about decreased choices and potential increases in traffic.

China

Beijing 2035[4]

Beijing released the general city plan for 2016 to 2035 accompanied by an objective to become a "world-class harmonious and liveable city" with no exact keyword of smart growth in its main goal or title. All the same, this plan has two principal goals of eliminating non-capital functions and solving "big city diseases". The city must take advantage of integrated development of the Beijing, Tianjin and Hebei regions, as well as optimization of capital, to be considered a national centre of politics, culture, international exchanges and scientific and technological innovation in order to achieve this. The latter, which deals with the blueprint, for example, has an option to re-establish the plan of the city into a central city area, a sub-centre, two axes and ten suburban areas.

This layout predicted a population of about 23 million residents. The ground should be decreased to about 2,860 sq km and 2,760 sq km for construction by 2020 and 2035, respectively. Thanks to green vehicles in the Beijing 2020 plan, pedestrians may not walk more than ten minutes through downtown to arrive at a subway station, breathing in fresher air. However, Beijing transportation authorities represent the blueprint in this way to interpret the capital's transportation development during the period of 2016–2020. The metropolis Beijing 2020 plan, with a projected population of over 21 million, will extend its 554-km urban rail network to more than 900 km.

The 1,000-km bus lane network will provide faster travel by developing new energy-efficient buses which up to 65 percent of the all buses will use that new energy plan by 2020. The plan will call for more bike transportation by supplying 3,200 km of bike lanes and at least 100,000 bicycles for rent. The public transportation system has a plan to link and integrate the roads of Beijing to neighbouring Tianjin Municipality and Hebei Province, providing commuters with travelling routes between the three regions with just a single public transportation card. Beijing has a plan to improve the road network in its eastern suburb of Tongzhou, which offers a "subsidiary administrative center", and the modification of road conditions in rural districts. In July 2016, Beijing announced it would move some administrative issues to the Tongzhou District by 2017.

Shanghai 2050[5]

The Shanghai city strategy for 2050 appoints Shanghai as the core city of the Yangtze River Delta city cluster and the capital of finance, international economy, trade, shipping and technological innovation. This objective might be realized by striking a balance between preserving agricultural lands and improving urban spaces, adding to the quality of life by allowing all ages to enjoy life and lead healthy lifestyles and firmly control the population.

Shanghai has to evolve multifarious strategies to set itself apart as a global city in the future with the help of both short- and long-term plans designed to satisfy the requirements of the growing urban population. The vast majority of issues, including employment, education, transportation, energy, housing, infrastructure and healthcare, are referred in this master plan, which focuses on human–centred design (HCD). This layout visualizes a city with a population of 25 million permanent residents by 2035, while the overall construction will be limited to 3,200 sq km.

Shanghai shall comply with smart growth strategies such as mixed-use land policies, planned densification and transit-oriented development (TOD) in order to support the complete utilization of areas by 2030. Shanghai can utilize the space and land more efficiently with the help of the HCD policies through comprehensive progress and strict tightening of regulations in terms of land size and use.

Public transport is a priority for Shanghai's urban planners. The rail transportation, for example, was strengthened by better integrating rail with highways, using deep underground spaces and improving transfer nodes. The policies also call for green commuting by considering separate motor and non-motor lanes as well as special bike paths. In addition to the reduction in traffic congestion and commuting times, HCD policies in TOD could offer innovative employment opportunities to neighbourhood residents.

Shanghai is estimated to experience a population growth from 27 to 34 million by 2050. This might apply pressure on both resources and the environment while the urban development plans are undertaken, which may cause reduced consumption of natural resources and space, controllable growth and sustainable development.

Theories of green technology and clean energy were applied in the plans and strategies of urbanization, such as open public spaces to create low-carbon communities. About 3,500 square kilometers of protected "ecological land" are included in the urban plans, which mainly provide economic development for eco-tourism. Shanghai's coastal areas have enacted disaster resilience prevention efforts, including a smart floodwall. Household wealth would be increased in Shanghai by carrying out energy-efficient policies and conserving open-space areas while improving environmental quality. The environmental HCD will eventually limit future resource bottlenecks to zero-growth consumption by 2040.

The economic competitiveness in terms of promotion and sustainable development in Shanghai will help facilitate its global aspiration over the next 35 years. Furthermore, with a focus on HCD principles in urban planning and community development efforts for urbanization, Shanghai can be compared to cities in the United States. They are also focusing on smart growth strategies to address sustainable development by making strong communities with affordable housing. An exchange of policy between the United States and China on innovating and transforming urban planning and HCD can expand global efforts to cooperate and promote strong communities supported by sustainable development.

India

New Delhi[6]

The population of New Delhi and its neighboring suburbs has doubled from approximately 10 million in 1991 to 22 million in 2011. The construction of mass rapid transport networks was not a barrier to the doubling of the number of private motor vehicle trips during 2001 to 2008, so congestion increased and the city was known as the "world's most polluted city".

Planners launched an overall investigation into the regulations by recognizing the major transformation requirements concerning urban development. Authorities also realized the need for integrating land use and transport planning while discouraging the use of private vehicles. In the end, the Delhi Development Authority (DDA) approved a new transit-oriented development policy in early 2015.

Higher densities within 500 m, or a ten-minute walk, of rapid transit stations are supported by provisions at the core of the policy. These areas are called "influence zones" in the Delhi Master Plan for 2021. The objective is to reduce the distance of trips by enabling a greater portion of the population to live in these influence zones. Higher floor-area ratios (FAR) – up to 400 percent of the plot area – are now allowed for redevelopment projects larger than 1 ha. The policy has also issued mixed use in developments in influence zones, with a minimum of 30 percent floor space reserved for residential uses, 10 percent for commercial uses and 10 percent for community purposes. This mixed-use approach is expected to restrict using a private vehicle for daily errands, and it helps create a higher concentration of jobs and residences within easy reach of rapid transit.

TOD is not just about higher density. An appropriate urban design can transform Delhi from being a "rape city" to a "safe city" by creating a better public–private interface. Based on these policies, an active interface between activities inside the buildings and on the street is assured by removing obstacles in front of building facades and authorizing transparent fences in front of allowed obstacles.

In addition, the strategy intends to enhance the public realm through people-centric street design guidelines. At least five crossing opportunities for each kilometer of street length in wide footpaths would be provided for pedestrians. Twenty-one percent of the land area is already used for roads in New Delhi. With these street design guidelines, its TOD policy looks at ways to better manage the existing road network to balance the needs of all users.

Encouraging development around transit can lead to disincentives for the use of private motor vehicles. Public transport in New Delhi is not all that desirable due to the presence of unrestricted, cheap on- and off-street parking. This can be addressed by considering the fact that the master plan eliminates the parking supply by allowing a maximum of 1.33 equivalent car spaces (ECS) for every 100 sq m of built space in TOD areas – far lower than the parking permitted in other areas poorly served by public transport. A small portion of this parking is used for cars. Provision of cycle parking (for bicycles, scooters, hoverboard, etc.) is mandatory in all developments. Some other forms of paid, publicly accessible and shared parking can be built.

The TOD approach in New Delhi was not fully comprehensible by many city planners initially. The attainability of a physical infrastructure – including electricity, water, sewage and solid waste management – was reviewed to support higher densities. For this purpose, the DDA, with the help of planners, urban designers and infrastructure experts, tested TOD concepts in a pilot project at Karkardooma. This analysis resulted in a high population density to be accomplished in association with social, ecological and economic viability. The findings also revealed that developments can easily be designed for mixed-income groups, with various compatible uses and with a decentralized infrastructure for water, solid waste and electricity.

The DDA has adopted a public–private participation (PPP) approach to speed up project implementation and support needed infrastructure investments.

Mumbai's smart growth plan[7]

About four centuries ago, seven islands inhabited by local fishermen together formed the city of Mumbai. Eventually, it became the trade and commerce capital of the country. By creating an island city mainly through large land reclamation and a large support infrastructure, including water supply and transport, this characteristic (trade and commerce capital for the country) was reinforced throughout the period of colonial rule. Now, within the jurisdiction of the Municipal Corporation of Greater Mumbai (MCGM), Mumbai encompasses 432 sq km and the Mumbai Metropolitan Area (MMA) encompasses 4,500 sq km, including 16 small and medium municipal corporations, including Mumbai itself.

Mumbai saw a rapid population growth for quite a long time due to trade and commerce, the growth rate of which descended to a modest level only in the last decade. The population trend in terms of size and structure during 1971–2011 was initially bound to the island city of Greater Mumbai. Suburbanization took place 20 to 25 years ago, while successful industrialization of the textile industry on the periphery of the island city and the suburbs formed the backbone of its economy, helped by the development of the existing rail network. Furthermore, the population growth of Mumbai was mainly taking place in the suburban areas and beyond the city borders. Early 2000 was an important turning point when the "deindustrialization" and "deconcentrating" of the island city and near the suburbs took place with the second development strategy of Mumbai.

Mumbai's master plan for 2034 adopted a responsive and flexible plan yet controlled development by readjusting its vertical space with horizontal spread. This will shift the population growth to the island and central parts of the city with a more proper infrastructure. The old industrial areas which are abandoned now and the old market yards/warehouses which lie vacant now are potentially appropriate for leveraging this growth, or smart growth. This plan promotes dense city centres and allows infill development to contain the growth outside the city. The current strategy as articulated in the regional policy to develop several city centres, or central business districts (CBDs), is difficult to adopt under the present regulatory regime, which poses several coordination issues. The suburban trains are the mainstay of the life of Mumbai, but they handle too high a load in the city peripheral areas rather than in the city centre.

The new policy has allowed the floor space index (FSI) levels to be changed to 3 for all kinds of residential construction in the island city. FSI, also known as FAR, is a tool to define the extent of construction permissible on a plot. The existing development plan of Mumbai formulated in 1991 had calculated the FSI levels in the island city at 1.33. While the previous Congress–NCP government had later allowed additional FSI on payment of premiums in the suburbs and the extended suburbs of Mumbai, the benefit wasn't extended to parts of the island city. As long as the new policy has retained the zonal (basic) FSI of 1.33, it has allowed an additional FSI up to 1.67 to be loaded in the form of paid FSI, either by paying premiums to the municipality or purchasing transferable development rights (TDR) from the open market, or a combination of both.

This is also the first time since 1991 that the permissible FSI levels in the island city have been raised higher than those in the suburbs (up to 2.5). It can be noted that real estate rates in south and central Mumbai are the highest in the country. The Fadnavis government first allowed paid FSI in the island city by permitting utilization of the TDR in November 2016.

Government planners have also adopted a co-living and working formula of sorts, which is expected to have a maximum impact in the island city. The government has also announced the allowed FSI levels for commercial developments is now 5. Despite the applicability of that in Mumbai, urban planners believe that the immediate beneficiaries of the move may well be developed commercial business districts and retail hubs in south and central Mumbai. Development of the co-living and working concept led urban planners to state that "residential construction up to 30 percent of the permissible FSI can be undertaken in such commercial developments". The main goal of increasing commercial FSI regarding the development plan document is to "reassert Mumbai's primacy as India's commercial capital" and create 23 lakhs more jobs in the city. Government planners assume that the residential construction condition will gain traction, especially among the younger lot.

Iran

Bojnourd[8]

Bojnourd, the capital of the North Khorasan province, with an area of 36 square kilometres (Figure 3.2), is located in the northeast of Iran. The initial urban development plan for the city was prepared in 2011 on the basis of smart growth. For the following four reasons, smart growth has been selected as the main approach in this project:

1   Bojnourd is located inside the bowl-shaped valley created by fault lines. Thus, the continuous physical development of the city is only possible within the plain, and the expansion of the city towards the plain edges causes it to grow on fault lines, which is inappropriate.
2   Bojnourd is surrounded by very fertile agricultural lands. Therefore, given the ecological indicators and the theory of sustainable land development, continuous physical development is not possible.
3   From a geological viewpoint, except for the areas limited to the city, the soil of the Bojnourd plain does not allow for physical development of the city, because a large part of the plain is calcareous, due to the dissolution of lime,

FIGURE 3.2   Bojnourd surrounded by mountains

Source: Retrieved from Google Maps in 2018

which because of moisture and consequently corrosion and erosion of the soil, means structure could possibly collapse. Any urban construction on this terrain faces serious financial and life-threatening risks.

4   In terms of topography, most of the Bojnourd plain is made up of high mountains with a steep slope, and thus, it has limited urban development on these steep heights, because the ground slope facilitates soil erosion, and even in some cases, slopes of more than 30 percent can be considered as impossible to walk and a negative factor. Slopes suitable for urban development are only up to 9 percent. In terms of ground stability/instability, slopes of less than 5 percent are considered sustainable for development, but on slopes greater than 5 percent, depending on the rock materials and structure, instability come on gradually.

In the Bojnourd smart development plan, the current situation and the progressive development of the city have been investigated from physical, demographic, social and economic viewpoints. The growth rate, urban development model and estimation of the housing needs for the whole city and the socioeconomic dimensions in urban neighbourhoods were studied. On the one hand, the goal is to adopt useful deterrent policies and objectives by recognizing the disproportionate development of the city, its population and the form of urban development, and on the other hand, recognizing the socio-economic dimensions, the necessary conditions and the context is needed to overcome the shortcomings of neighbourhoods, provide useful service delivery, modify contradictory land uses, improve the economic situation, increase the social relations and, ultimately, improve the living conditions of the inhabitants. In the Bojnurd Urban Development Plan, the following policies have been proposed with a view to smart growth:

1 Adopting policies that deter horizontal growth and promote the city's demographic and physical development in the form of vertical development, especially in Nazerabad;
2 Prioritization of development and construction and the use of incentive policies for vertical constructions and centralized urban development;
3 New strategies for managing property and preventing land and housing speculation, leading to phenomena such as lands left unconstructed within the urban fabric or in sprawled fabric and unoccupied buildings;
4 Improvement, renovation and reconstruction of residential worn-out urban textures;
5 Prioritization of pedestrians over those riding, and creating streets devoted to walking with special recreational-commercial attractions that pave the way for social activities as well as the economic growth of the neighbourhoods;
6 Create mixed-use neighbourhoods by improving the quality and quantity of existing services or applications or building new applications in accordance with the priority of the average service delivery of applications and standardized scores in each neighbourhood;
7 Attention to all land uses, especially recreational, green space, educational and healthcare-medical, in Nazerabad and completing the construction of the only mosque located in the same neighbourhood
8 Planning for the construction of new residential units and urban infrastructure, given the estimated housing and land needed for residential and non-residential uses in the future;
9 Emergency and long-term planning for educational purposes and employment in Khavar Mahale, as well as shopping and employment in other neighbourhoods, given the high volume of intra-city trips, aiming to access these services in each neighbourhood;
10 Building a diverse range of residential units with the greatest diversity in terms of income groups, education levels and ethnicities;
11 Adopting appropriate measures to increase walking and to reduce the use of personal cars, considering the highest use of personal cars and the largest share of the income spent on urban trips;
12 Planning to improve the flooring, bonification, safety and ease of movement on the sidewalks in Khavar Mahale, given the high volume of walking trips from other neighbourhoods to this district;
13 Planning to build sidewalks in Nazerabad, given that the alleys of this neighbourhood lack appropriate sidewalks;
14 Creating complete and integrated streets suitable for the traffic of pedestrians, cyclists, and drivers;
15 Strengthening the sense of place and identity at the neighbourhood level, especially in Nazerabad, by improving social and economic factors.

Kerman[9]

Kerman is one of the largest metropolises in Iran and the capital of Kerman Province, the largest province in Iran, located to the southeast. According to the 2016 census, the city population was 738,724 people. Kerman is one of five historic cities in Iran. Kerman has an area of about 13,000 hectares, and due to its size and population, the city is classified as one of the major metropolises in Iran. Marginalization is the most important problem in the metropolitan area. A smart development plan on an urban scale has been developed for the city. The suggestions differ according to the studied area, because planning in each of these areas takes place in accordance with the facilities of the same area.

Within urban areas, the emphasis is placed on the redevelopment and intra-fabric development of existing neighbourhoods, increasing the combination of land uses and increasing the densities and more diverse transportation systems, especially walking and public transportation. Due to the size of Kerman, one can feel a need for smart urban development. There are many undeveloped and less developed areas in the municipal districts of Kerman. This lack of development is due to the imbalances between the four municipal districts in terms of land uses and their diversity to which one should pay attention in some districts in particular, like the fourth district; however, Kerman city is robust in terms of residential land use. Thus, the suggestions raised in the scope of this study are within the range of the combined land uses and density indicators, which are two basic elements that all planning considers. Therefore, considering the opportunities and potential available in each district, and with the help and guidance of other disciplines and techniques, steps should be taken to realize the solutions and suggestions for smart development in the urban districts of Kerman. The districts that are more robust in terms of commercial land use should be more focused on demographics, transportation, traffic and environmental issues. The sprawling expansion of the urban districts and the numerous economic and environmental impacts should cause the urban experts to address smart development policies to remove constraints, reach balance and equality, improve economic efficiency, reduce the additional costs and environmental issues and provide pedestrians access to narrow streets that reduce traffic, which is in contrast to "dispersion" and can be used as a reliable tool.

The study and results of the Kerman Smart Development Plan indicate that due to the existing physical and environmental issues, the plan requires a combined viewpoint (the combination of native knowledge and people's participation with the techniques and principles of smart development). In fact, the physical model of the city is not synchronized and in line with the population and the distribution of urban infrastructure and facilities. Urban districts experience imbalances among themselves in the distribution of land use, the heterogeneity and similarity of land uses and their combined diversity so that the centres of the districts, especially in the new urban textures (regions 1 and 2), are more developed than the other districts. Therefore, in the scientific and administrative realm, compaction must begin from within the urban centres according to the characteristics of each region and continue to build in these areas because the centres and their surrounding areas are prone to increased density and the construction of high-rise buildings. In the old texture, the city is faced with high density and a shortage of required service delivery land uses, and residential spaces are characterized by excessive physical damage and need to be improved. In the outer textures, failure to comply with construction plans and rules and regulations, the shortage of the land required by residential areas, sprawling construction, the undesirable conditions of quantitative and qualitative indicators and mixed land uses are quite evident and reflect the need for guidance and control. Socio-demographic studies of the districts do not follow a balanced and proportional distribution process, and the demographic and construction densities are not proportional to the physical development trend of the city; in addition, the sprawling physical development of the city has outpaced population growth. Environmental studies also indicate that in the urban centres and their surrounding areas, pollution is higher than elsewhere, and there is little green space available within the districts – in this regard, urban planning does not match the projections and per capita land allocation to all land uses, and this often leads to rework in the implementation process of the plans. For this reason, one should have a combined viewpoint of the process for growth and development within the city.

In general, despite the adoption of smart development in some countries and its success, using it as a long-term strategy in organizing the urban areas of Iran, using Kerman as a specific case, will have favourable results only when it can be of great help to promote urban development methods with respect to each area and region and is proportional to changes in attitudes and lifestyles over time and with respect to spatial differences. In the following, the practical strategies of the Kerman Smart Development Plan are examined in terms of emergency, short-, medium- and long-term scales.

Emergency suggestions:

1   The attention of urban management authorities and planners to the compact city model;
2   Preparing the compression plan as the most important principle;
3   Estimation and evaluation of facilities of each region, as well as financial resources and facilities;
4   Planning for the formation of a team and group of experts in other specialties and in various disciplines in the course of the project and taking advantage of the views of all of them;
5   Estimation of the population of the districts and the types of land uses within them, as well as the shortages;
6   Determining the more developed and less developed regions;
7   Districts need to make the most important issues a priority.

Short-term suggestions:

1   Optimal use of deserted spaces and abandoned lands;
2   Reconstruction and renewal of old regions;

3   Planning in order to provide a variety of transportation alternatives (bicycles and pedestrian-oriented options);
4   Encouraging communities and people to participate in development;
5   Increasing the construction density in less developed regions;
6   Creating green spaces, children's playgrounds and local parks.

Mid-term suggestions:

1   Planning and creating mixed and diverse land uses.
2   More control over urban areas: One of the main causes of sprawl is the lack of attention and planning for the development and expansion of the region by the relevant organizations and planners or their failure to implement the relevant plans. Therefore, urban organizations such as the municipality should avoid unplanned development of the region, which is often raised by land traders and speculators.
3   Construction with mixed land uses on a scale suitable for districts and neighbourhoods.
4   Creating opportunities to change single-user retail and commercial construction into communities with hybrid and mixed walkable land use.
5   Creating financial incentives to help in the implementation of urban smart growth projects.

Long-term suggestions:

1   The availability of streets and alleys providing access for pedestrians, and only connect to the main roads, and contribute to the vehicle traffic flow;
2   The presence of landscapes near the sidewalks because people observe close spaces in this compact form of a city;
3   The creation of a balance between environment and development;
4   Increased accessibility to the services (such as green spaces, education facilities, etc.) of low-income households;
5   Reducing physical distances between residential area and important nudes such as grocery shop, elementary school;
6   Diversity of transportation opportunities and reduction of urban traffic;
7   The systematic, planned and sustainable development of the city.

Discussion and conclusion

To summarize global experiences, the principles, objectives and perspectives of practical examples examined in this chapter were compared to the ten-plus-one principles of smart growth. As can be seen in Table 3.5, many of these principles can be traced in the practical cases mentioned in this study. Although Chapter 4 addresses the critiques, challenges and obstacles to the realization of smart growth in detail, this chapter also attempted to explain and summarize the benefits and

TABLE 3.5 Comparison of the principles, objectives and perspectives of global experiences with the ten principles plus additional principle of smart growth

Smart growth in the state/city	Principles/objectives/perspectives	Ten principles of smart growth
Idaho	Combined transport	Principle 8. Providing a variety of transportation choices
	Mixed uses	Principle 1. Mixing land uses
	Housing	Principle 3. The provision of various housing choices
	Sense of place and social appeal	Principle 5. Fostering distinctive, attractive communities with a strong sense of place
	Pedestrian-oriented neighbourhoods	Principle 4. Creating walkable neighbourhoods
California	Land use	Principle 7. Strengthening and directing development towards existing communities (infill development)
	Housing and employment	Principle 3. The provision of various housing choices
		Principle 9. Making development decisions predictable, fair and cost-effective
	Public transport	Principle 8. Providing a Variety of Transportation Choices
	Infrastructure	
Maryland	Population density	Principle 2. Compact development
		Principle 7. Strengthening and directing development towards existing communities (infill development)
	Jobs/housing balance	Principle 3. The provision of various housing choices
		Principle 9. Making development decisions predictable, fair and cost-effective
	Mixed uses	Principle 1. Mixing land uses
	Street network pattern	Principle 8. Providing a variety of transportation choices
Ontario	Cheap housing	Principle 3. The provision of various housing choices
		Principle 9. Making development decisions predictable, fair and cost-effective
	Promoting sustainable transportation	Principle 8. Providing a variety of transportation choices
	Dense city centres	Principle 7. Strengthening and directing development towards existing communities (infill development)

Smart growth in the state/city	Principles/objectives/perspectives	Ten principles of smart growth
	Preserving land resources	Principle 6. Preserving open space, farmland, natural beauty and critical environmental areas
Portland	Safe and stable neighbourhood	Principle 4. Creating walkable neighbourhoods
	Compact development Healthy economy	Principle 2. Compact development
	Protection of environment	Principle 6. Preserving open space, farmland, natural beauty and critical environmental areas
	Balanced transportation	Principle 8. Providing a variety of transportation choices
	Housing for all	Principle 3. The provision of various housing choices
San Diego	Preserving natural resources	Principle 6. Preserving open space, farmland, natural beauty and critical environmental areas
	Limiting urban sprawl	Principle 2. Compact development
	Housing planning	Principle 3. The provision of various housing choices
	Regional and local planning	Additional principle: The need for collaboration among different players
	Comprehensive Transportation	Principle 8. Providing a variety of transportation choices
	Environmental Mitigation Program	Principle 6. Preserving open space, farmland, natural beauty and critical environmental areas
	Collaboration and incentives	Principle 9. Making development decisions predictable, fair and cost-effective
Beijing	The opportunity of coordinated development	Principle 9. Making development decisions predictable, fair and cost-effective
	Big city diseases	Principle 2. Compact development
	Limiting land construction	Principle 2. Compact development
	Walkable neighbourhoods	Principle 4. Creating walkable neighbourhoods
	Various transportation modes	Principle 8. Providing a variety of transportation choices
Shanghai	Saving agricultural land	Principle 6. Preserving open space, farmland, natural beauty and critical environmental areas
	Controlling boundaries	Principle 2. Compact development

(Continued)

TABLE 3.5 (Continued)

Smart growth in the state/city	Principles/objectives/perspectives	Ten principles of smart growth
	Human-centred design	
	Public transport	Principle 8. Providing a variety of transportation choices
	Strong communities	Principle 3. The provision of various housing choices
	Regional decisions	Additional principle: The need for collaboration among different players
New Delhi	Transit-oriented development	Principle 8. Providing a variety of transportation choices
	Mixed use	Principle 1. Mixing land uses
	Public–private interface	Principle 9. Making development decisions predictable, fair and cost-effective
	Higher densities	Principle 2. Compact development
Mumbai	Infill development	Principle 7. Strengthening and directing development towards existing communities (infill development)
	Regional policymaking	Additional principle: The need for collaboration among different players
	Using vertical space	Principle 2. Compact development
	Connected urban transport	Principle 8. Providing a variety of transportation choices
Boujnord	Vertical construction	Principle 2. Compact development
	Walkable streets	Principle 4. Creating walkable neighbourhoods
	Mixed-use neighbourhoods	Principle 1. Mixing land uses
	Variety of houses	Principle 3. The provision of various housing choices
	Complete streets	Principle 8. Providing a variety of transportation choices
	Improving sense of place	Principle 5. Fostering distinctive, attractive communities with a strong sense of place
Kerman	Infill development	Principle 7. Strengthening and directing development towards existing communities (infill development)
	Expanding transportation choices	Principle 8. Providing a variety of transportation choices
	Encouraging public participation	Additional principle: The need for collaboration among different players
	Walkable neighbourhoods	Principle 4. Creating walkable neighbourhoods
	Mixed use	Principle 1. Mixing land uses

challenges of smart growth based on global experiences, along with the principles, goals and criteria required for achieving it.

In fact, one of the most important goals of examining global experiences is to investigate the reasons behind the approaches taken to reach smart growth in North America and other regions as a platform for the formation of this urban concept. Based on what was mentioned in this chapter, it can be said that the smart growth plans implemented have formulated their approach based on the strengths of the city or attempts to cover existing weaknesses (e.g. the emphasis of the Garden City Plan on affordable housing). A remarkable point that has become more tangible in this chapter is the fact that the nature of smart growth has been able to cover a wide range of urban issues due to its deep perspective.

Smart growth in the urban plans of Idaho follows five different processes. In this way, smart growth shows a high degree of flexibility and can change its focal point in accordance with the conditions. Therefore, it should not be assumed to be a fixed form or a module strictly applied to various conditions.

At the beginning of this chapter, the implementation of smart growth in Idaho was investigated. The first goal in planning for the cities of this state was to increase transportation options for citizens. The second goal was to increase use mixing and reduce the workplace–housing distance for residents, which alone could address urban issues such as traffic congestion, conservation of natural and environmental resources, readability of the urban form, etc. Other goals of these projects include the production of affordable housing and the creation of a sense of place. These goals are remarkably similar to those for Riverside, California. The four main topics mentioned in California's Riverside plan include land use (use mixing in the Harris Ranch community), housing and employment (the Garden City plan), public transportation (the Ada Urban Areas Highway District) and infrastructure (the Sandpoint plan) – these were also featured in the Idaho's smart growth program. The most significant phenomenon in California's Riverside program is that of land use. This sometimes has affected the size of land lots (in the form of a compact development indicator), housing affordability (various housing options) and elsewhere mixing uses. In this way, it may be argued that the discussion of land and its use is the most important aspect of smart growth for both California and Idaho. It can be argued that the dominant form of urbanization in these two states has been shaped and developed on the basis of open lands with villas, with an emphasis on the swinging relationship between city centres and suburbs. Such urban expansion, whether in urban form and morphology or wider spatial planning, is in conflict with smart growth principles. Thus, the point of emphasizing land use is to resolve a problem that has deprived the city of its vitality, a point that is supported by examining Maryland's experience in smart growth–based design. Maryland determined criteria such as demographics, population density, use mixing, residential density, etc. based on land-based computer modelling. The modelling software used in the smart growth design of Fells Point was a quantum GIS derivative that examined various scenarios in the form of "if–then" to plan land use in the area under consideration.

In China, two major cities, Beijing and Shanghai, and their master plans were discussed briefly. Both of these cities tried to control population in order to limit the city boundaries. Shanghai used human-based design as the prime core of its plan in order to create sufficient public services and a smart city. Similar to Beijing and Shanghai, New Delhi and Mumbai are two major cities in India with massive populations. New Delhi used TOD as its main approach to designing a future transportation plan, and its goal is to create a safe city for pedestrians. As mentioned before, private cars are the dominant form of transportation, and the city master plans tried to balance the needs of all users (private and public transport). In Iran's experience, the first case of Bojnourd has natural limitations which made smart growth the number-one solution for the city's future development. Kerman's plan for smart growth emphasizes multi-scale and plural management to create a better future. Although Kerman's plan has the same principles of other smart growth plans, classifying its strategies into four time-consuming categories is the main difference from other cities' smart growth plans.

This chapter bridges theory and practice and aims to identify solutions used to achieve smart growth in urban development. It seems that land use is the main topic of consideration in the United States. In China and India, the main topic is controlling overcrowding of their cities along with providing a sufficient transportation system. In additional, reviewing these cases shows that besides the ten principles introduced by the Smart Growth Network, many other actions embody the goals and targets of smart growth. For instance, the principle of regeneration or revitalization of old downtown areas can be added, or having a wider perspective that analyses the city within a wider regional area and could lead to better planning of a transit system from the outside to the inside of a city. As another example, healthy economies should be added to smart growth principles as a key factor which could improve the quality of life of citizens, along with a human-scale urban space. Reviewing the global experiences illustrates that smart growth is not a one-size-fits-all or uncompromising guideline that must be used for each city; rather, it's a starting place for people to find a suitable smart plan for their cities.

Notes

1 The ideal scenario was developed by Maryland planners using the smart growth guide support data and other published sources.
2 This section is based on the Portland 2040 Plan for Growth (Oregon Metro, 2014).
3 This section is based on the San Diego Regional Comprehensive Plan for 2050 (SANDAG, 2018).
4 This section is based on analysis of the Beijing Master Plan for 2035 (Xinhua, 2017).
5 This section is based on the master plan for 2050 for Shanghai (Shanghai Government, 2018).
6 This section is based on analysis of the Delhi plan for smart growth around transit (Institute for Transportation and Development Policy, 2015).
7 This section is based on an analysis of *The Indian Express* on the Mumbai master plan (Ashar, 2018).
8 This section is based on the Bojnourd smart growth plan for sustainable development (Bojnourd Municipality, 2011).
9 This section is based on the Kerman City Master Plan (Smart Development) for 2009 to 2034 (Kerman Municipality, 2009).

References

Ashar, S. A. (2018). *New masterplan of Mumbai: Repopulate the island city*. Retrieved from India, https://indianexpress.com/article/cities/mumbai/new-masterplan-of-mumbai-repopulate-the-island-city-development-plan-2034-5152049/

Blais, P. (2011). *Perverse cities: Hidden subsidies, wonky policy, and urban sprawl*. Vancouver: University of British Columbia Press.

Boise City Council. (2006). *Harris ranch specific plan*. Retrieved from USA.

Bojnourd Municipality. (2011). *Smart growth plan of Bojnourd in order to achieve sustainable developments*. Retrieved from Bojnourd.

Burchell, R., Downs, A., McCann, B., & Mukherji, S. (2005). *Sprawl costs: Economic impacts of unchecked development*. Washington, DC: Island Press.

Clegg, E. (2008). *Putting smart growth policy into practice*. Retrieved from USA.

Colliers International. (2007). *City of meridian – ten-mile specific plan*. Retrieved from USA.

Eidelman, G. (2010). Managing urban sprawl in Ontario: Good policy or good politics? *Politics & Policy, 38*(6), 1211–1236.

Hogle-Ireland. (2007). *Western riverside smart growth opportunity*. Retrieved from USA.

Idaho Smart Growth. (2014). *Smart growth best practices*. Boise: Idaho Smart Growth. Retrieved from http://www.idahosmartgrowth.org/app/uploads/2014/05/smart_growth_best_practices_21.pdf

Institute for Transportation and Development Policy. (2015). *Delhi plans for smart growth around transit*. Retrieved from www.itdp.in/delhi-plans-for-smart-growth-around-transit/

Kerman Municipality. (2009). *Kerman city smart development*. Retrieved from Kerman.

Ministry of Infrastructure. (2011). Places to grow. Retrieved from www.placestogrow.ca/index.php?option=com_content&task=view&id=281&Itemid=84

Ministry of Municipal Affairs and Housing. (2015). Future development up to 2031 under the Growth Plan for the Greater Golden Horseshoe Ministry of Municipal Affairs and Housing.

Office of Smart Growth. (2005). *Smart growth: Maryland*. Retrieved from USA.

Oregon Metro. (2014). *2040 Growth concept*. Retrieved from Portland, www.oregonmetro.gov/2040-growth-concept

Ruth, M. (2006). *Smart growth and climate change: Regional development, infrastructure and adaptation*. Cheltenham: Edward Elgar Publishing Limited.

SANDAG. (2018). *San Diego regional comprehensive plan 2035*. Retrieved from California, www.sandag.org/index.asp?classid=12&fuseaction=home.classhome

Sewell, J. (2009). *The shape of the suburbs: Understanding Toronto's sprawl*. Toronto: University of Toronto Press.

Shanghai Government. (2018). *Master plan for Shanghai 2050*. Retrieved from Shanghai, www.shanghai.gov.cn/newshanghai/xxgkfj/2035004.pdf

Smart Growth BC. (2008). *2008 Annual report*. Retrieved from Vancouver.

The Transpo Group. (2005). *Pedestrian-bicycle transition plan*. Retrieved from USA.

Tomalty, R., & Alexander, D. (2005). *Smart growth in Canada: Implementation of a planning concept*. Ottawa: CMHC.

White, R. (2007). *The growth plan for the Greater Golden Horseshoe in historical perspective*. Toronto: Neptis Foundation.

Xinhua. (2017). *China focus: China sets population, construction limits in Beijing city planning*. Retrieved from Beijing, http://en.people.cn/n3/2017/0928/c90000-9274639.html

4

A REVIEW OF THE CRITIQUES OF SMART GROWTH

Introduction

Over the past three decades, smart growth has experienced tremendous development in the United States in terms of strength and influence, both as a movement and as a set of values and ideas (Liberty, 2004). However, in the current situation, although many experts agree with such a development, there are still contradictory feelings about the exact position of smart growth. From the point of view of many world-renowned scholars, in recent decades, smart growth as an influential urban movement has been able to promote important changes in urban growth and performance. Nevertheless, not everyone considers smart growth to be as important as what some of its supporters believe. Some also believe that this approach still does not receive the attention it deserves (Theart, 2007). Despite significant discussions about the benefits of smart growth, critics contend that smart growth does not have significant benefits to cities, but instead imposes significant costs on the urban system. Some critics also find that smart growth's ability to correctly manage the urban expansion is lacking and believe that smart growth exacerbates various aspects of density and, as a result, worsens the conditions of the urban population. They argue that smart growth increases traffic congestion, air pollution, accidents, public service charges, crime and poverty (Ziari, Hatami Nejad, & Turkmennya, 2012).

In a report for Victoria Transport Policy Institute in 2011, Litman summarizes the most important criticisms regarding smart growth in the following areas:

- The increase in legal rules and regulations and decrease in social freedom;
- The increase in housing costs by reducing the use of land;
- Densification and reduced public transportation quality;
- The increase in public service costs (Litman, 2011).

According to this analysis, Litman argues that though critics of smart growth point to some logical flaws in his assessments, they also make false conclusions and draw a false image of smart growth. For example, smart growth evaluations are based solely on gross regional population density and do not consider factors like the relationship between city size, density, crowding and transport patterns, on the one hand, and the tendency to use smart growth policies in areas of fast economic and demographic growth and consideration of the benefits and costs caused by its implementation, on the other, which can change the outlook on the subject (Thompson, 2012).

Yang, quoted in Cox and Litman, wrote that in some cases, smart growth ignores its commitment to its principles. From the perspective of critics, smart growth has led to an increase in urban living expenses. It can also increase traffic congestion and pollution. And it ignores users' preferences regarding space and makes individuals not think of single-family houses and cars. Therefore, it may be argued that smart growth limits personal freedom (Yang, 2009).

Important criticisms of smart growth

To help with the practical implementation of a smart growth approach, this chapter will address and analyse some of the most critical operational criticisms of this approach one by one.

Lack of attention to residents' preferences and failure to realize smart growth policies socially and economically

According to some people, the concept of a compact city can be alluring because they believe that this approach has attracted attention due to the support of the general public, as well as its rapid transition to political issues. But another group of smart growth critics state that there are significant gaps in the environmental claims of a dense city. From their point of view, empirical analysis reveals that the environmental benefits of urban compression policies may not be to the extent that they are claimed. At the same time, special attention should be paid to the economic and social costs of this approach. From the critics' point of view, there has always been a tendency to concentrate on the pioneering role of planning and design in developing a compact urban plan, but less attention has been paid to the social, economic and technical processes involved in shaping the feasibility and realization of this concept (Rahnama & Abbas Zadeh, 2006).

Gordon and Richardson (2000) believe that the supporters of smart growth offer few analyses and discussions in terms of the future costs, implicit exchanges, possibility of the persistence of horizons or even public desire as consumers of this space in such communities. Instead of imposing things that only benefit the development, they also believe that the public must be given the right to choose. According to Gordon and Richardson (2001), the market has a better performance record than smart growth planners in this regard (Gordon & Richardson, 2000, 2001).

In other words, some smart growth critics believe that existing policies should be tailored to people's wants and be developed as neutral and impartial. According to these critics, people in automotive-driven modern societies prefer large, single-family houses. In fact, from the perspective of some of these people, the current transportation system and current land use policies are fair and efficient and there is no need for them to change (Bashiri, 2011). Based on this view, the current patterns of land use and transportation reflect the priorities of consumers. Hence, applying policy changes based on the smart growth approach is not desirable for consumers and is in contrast to market demand (Ziari et al., 2012).

Nonetheless, the point to be noted is that the preferences of people are different and often include some qualities that can be achieved in scattered cities and some that can be achieved by smart growth urban policies. For example, although market surveys indicate that most families prefer to reside in single-family houses, research also suggests that many people consider important factors such as accessibility and diversity of transportation systems. Therefore, it can be said that in many cases people tend to have a simultaneous mix of some of the qualities of the local communities in city centres combined with the benefits of the scattered urban context of the suburbs.

In order to examine the general public's preferences about and real interest in the suburbs, it is necessary to distinguish between the physical desirability of scattered development (such as lower congestion levels and car-dependent land use patterns) and socio-economic characteristics of dense development (which can be created in smart growth–based communities through urban renewal or the creation of more compact, multi-faceted suburbs). In fact, only a few of the characteristics mentioned by the public for their selection of the suburbs are physical features that cannot be replicated in other urban environments. There are many other physical wants, such as the desire to have large yards and large private green spaces, and these can be provided in urban environments through the sharing of green spaces (such as public parks) among several households.

At the same time, critics reject the relationship between city size and volume, density, travel pattern, income and living expenses, on the one hand, and the tendency for smart growth to be implemented in areas that are experiencing severe economic growth, on the other (Ziari et al., 2012). Also, they are not willing to accept deviations in the current market of land and housing that can lead to increased land use and motor vehicle trips. However, many smart growth strategies focus on land reform and housing fixes and are working to improve the financial conditions of consumers by increasing the efficiency of the existing economic infrastructure. Other strategies, such as tailor-made regulations and appropriate tax policies, as well as public investments to strengthen smart growth, can be used in conjunction with market reforms.

Lack of concrete effects of increasing public transport options to remove people's dependence on cars

Some critics claim that modern humans prefer scattered development and car-oriented communities. These critics believe that smart growth cannot properly respond to the needs of today's modern and busy families, who need to rely on car to meet their busy schedules (Litman, 2015). Critics emphasize that the idea of increasing public transport options in smart growth policies in order to reduce car travel is a waste of time, because most people deem travel by car the easiest option (Theart, 2007).

In fact, achieving a balance between pedestrian/bicycle lanes and car traffic has become one of the biggest challenges facing urban planners. In this regard, urban planners have to look at different options in order to find a way to increase the quality of footpath systems and other public transportation systems to a level where cars play a minor role in certain areas, especially in urban centres (Abrams & Ozdil, 2000).

In 1960, when cars were dominating cities, important decisions were made in the city of Copenhagen in Denmark. In order to reduce congestion and crowds, the authorities started converting streets into pedestrian walkways instead of expanding roads. The success of the decision led to the closure of more streets, with the reduction of parking areas in certain parts of the city. At the same time, the public transport system was upgraded and more cycling paths were built. Another positive effect of these actions, which took place a few decades ago, was that according to a public survey in major industrialized countries between 1973 and 1992, Denmark was the only country in which avoiding cars and switching to using bus and subway led to a dramatic reduction in energy consumption in travel and in emission of pollutants (Sheehan, 2001).

There are clear reasons indicating that many people in communities with smart growth choose other transportation methods, especially when supported by appropriate policies. In fact, this critique does not seem to be rational and proper, because many smart growth strategies help us save time (Rahnama & Abbas Zadeh, 2006). For example, smart growth increases accessibility, which in turn, shortens journey times, improves transport options and reduces the time needed for parents to take their children to school. Also, the conditions for cycling and walking improve in this way, and residents can spend time on exercise and recreation while traveling and going to work (Bashiri, 2011).

On the other hand, in the majority of cases, people living in suburban settlements have not been completely free in choosing their residences. Perhaps if their choices were not affected by factors such as the economic situation, the location of their workplace, etc. they might have preferred to live in dense urban contexts.

In fact, this group of critics does not consider many of the benefits of smart growth, such as the possibility of financial savings, increased physical mobility of individuals, community cohesion and better environmental protection.

The slowness of smart growth processes

A criticism that sometimes is posed against smart growth is that introducing changes in the land use process is time consuming, and its effects and benefits can only be achieved several years later. According to some studies conducted in many local communities, only 4 percent of the land is subject to construction or redevelopment annually. Therefore, it often takes a decade to generate significant regional impacts through development based on a smart growth approach. In response, it should be noted that many of these changes can provide benefits and, at the same time, leave durable effects. In short, smart growth provides a long-term legacy of greater access and increased liveability for the local community in the future (Litman, 2012).

Preventing the growth of urban suburbs

O'Niell (1999) believes that many consider smart growth a code word for lack of growth. However, the opposite is true. Smart growth should be seen as a concept that increases growth. The only remaining issues are location, scale and how it grows. The reality is that cities are rapidly growing. Therefore, instead of trying to restrict development, more emphasis needs to be placed on the development management method. From the perspective of O'Niell, smart growth is often seen as an anti-suburban approach. Therefore, some urban planners and developers only focus on the central parts of cities.

Instead, smart growth should be seen as a concept that seeks to bridge the gap between density and market-based development in urban and suburban areas (Theart, 2007). In fact, many smart urban development policies can be implemented in the suburbs in the form of new towns and cities and change their structure. Thus, cities will encounter suburban areas that are developed after the capacity of existing cities has been completed, and smart growth criteria, including physical density, have been considered in their development.

Enforcing strict compliance with rules and regulations

Some critics believe that smart growth policies act as just another layer of government regulations that slows down the urban development process. However, if the regulations and conditions are corrected as intended by the concept of smart growth, the processes will be simpler. This way, beneficial projects get the necessary permissions faster, and their implementation becomes facilitated. Many of the current rules and regulations around the world that are designed to create low densities differentiate various uses from residential areas, making it more difficult to implement the concept of smart growth. Smart growth supporters seek to modify construction guidelines and codes, zoning plans and uncontrollable regulations in order to simplify the methods and, in turn, make it easier to build projects reflecting smart growth characteristics.

Land allotment and limitation of environmental dynamics

Wendell Cox (2011), as an opponent of smart growth policies, points out that smart growth theory imposes land allotment and limits on environmental dynamics. To him, this does not reduce the problems of cities, but increases them.

He assumes that smart growth increases *prescriptive land use regulations*. In fact, it increases some but reduces many others. Although smart growth policies may include various regulations and incentives that discourage urban expansion, they reduce other regulations, including limits on building density, height, setbacks, mix and minimum parking requirements.

Cox and Utt (2004) compare smart growth and urban dispersion and argue that lower costs are not necessarily associated with dense cities, structural growth and growth based on existing cities, but can be achieved through the distribution of low and medium density in new cities with rapid growth.

In response to these critiques, it should be noted that smart growth principles do not allot lands and that the positive effects of the implementation of smart growth principles will emerge in the long run – it is not logical to expect a rapid decline in costs in a short time. Moreover, as pointed out, studies have shown that, in the short term, the application of smart growth principles will have a positive effect on reducing the demand for personal transportation and other aspects of inhabitants' functioning. Indeed, smart growth regulates developments and will have beneficial effects on environmental dynamics in the long run.

Restricting residential options and raising the cost of housing

According to the perspective of Gordon and Richardson (2000), acknowledging that urban development is inefficient and wasteful and that land and resources have been wasted, the supporters of smart growth are trying to make changes in these practices. On the other hand, the critics argue that limited construction and development in urban areas would reduce the housing costs and affordability for many community groups (Litman, 2011), and restriction on land supply would increase housing costs as well (USEPA, 2011). The study by O'Tool on *The Folly of Smart Growth* points out that whenever the principles of smart growth are applied to land use policies, as seen in Portland in the United States, the regional housing market is widely shaken (O'Tool, 2001). Theart also wrote that restrictions imposed by smart growth policies in 1989 led to the reduction of affordability of house purchasing for single parents (Theart, 2007).

In response to this critique, smart growth advocates state that it ignores the various ways in which household expenses are reduced. These include reducing the land needed to build each residential unit; increasing housing options; and reducing costs for installations, infrastructure, and transportation networks.

Increasing legal regulations and creating social constraints

Smart growth critics posit that smart development will largely increase legal standards and thus reduce individual freedoms. In other words, this kind of development is considered a kind of social engineering and a governmental repressive bureaucracy that can ignore and restrict the rights of real estate owners.

TABLE 4.1 Some of the most important smart growth limitations and capacities with an emphasis on individual and social freedoms

Increases freedoms	*Reduces freedoms*
– Allows higher density, more infill development	– Restricts urban expansion
– Allows more mixed land use	– Reduces traffic speeds
– Increases housing options (small lots, multi-family units)	– Increases parking fees
– Preserves existing neighbourhoods and communities	– Requires design standards and review
– Allows more flexible parking requirements	
– Reduced parking subsidies	
– Improved travel options, particularly for non-drivers (walking, cycling public transit, taxi services).	

Source: Litman, 2011

However, many of the smart development strategies have reduced existing regulations and increased the number of freedoms by creating various opportunities. In fact, smart development tends more to increase freedoms than to reduce them (HosseinZadeh Delir & Safari, 2012). For example, smart growth leads to an increase in individual and social freedoms by creating flexible plans and giving people the right to choose different travel options.

Another point to note is that social life is fraught with contradictions among various types of freedom. For example, the freedom of an individual to make noise violates the freedom of others to enjoy peace and quiet. There are some contradictions in growing suburbs, where more restrictive urban policies are used instead of rural land use policies with fewer constraints. Table 4.1 shows some of the most important smart growth effects on the creation of limitations and capacities and on individual and social freedoms.

Traffic densification and mixed urban uses

Critics claim that by increasing density, smart growth increases traffic congestion and air pollution (Litman, 2011, 2012). O'Tool (2001) also believes that smart growth policies only worsen the quality of life of urban residents because the increase in traffic congestion and air pollution increase the living expenses for the urban inhabitants and act as a barrier to urban livelihoods.

But a point that critics are unaware of is the fact that public transportation in dense areas is more efficient in terms of cost and speed of displacement, leading to fewer people using their own personal vehicles. On the other hand, mixing urban uses creates the opportunity for residents to work in the same localities as they live (Theart, 2007).

In fact, smart growth will provide more access and options to travel than it does an increase in traffic congestion. It should also be noted that traffic congestion cannot be considered an appropriate indicator for judging the quality of the transportation system (Ziari et al., 2012). If travel distances are reduced and effective ways of moving (such as walking and cycling) are advanced, less driving will be needed to reach destinations, resulting in less traffic congestion. Experimental data also indicate that smart growth does not increase the average time required for urban travel (USEPA, 2011).

Increasing urban density and social problems

Urban density is associated with social problems such as poverty, delinquency, prostitution and so on, and this is something stressed by some critics of smart growth. In fact, from the point of view of these critics, due to the constraints on the construction of multi-family housing in the suburbs and the reliance of these areas on car transportation, certain social groups are excluded from those areas, resulting in the concentration of poverty and social problems in neighbourhoods and urban contexts (Bashiri, 2011).

However, the important point here is that there is no scientific reason confirming that urban neighbourhoods cannot act as safe and dynamic places. Smart growth includes strategies that address many of these social problems and can reduce social problems by increasing social interactions and creating economic opportunities for urban residents. In fact, density and congestion, and the resulting increased access, can also have a variety of social and economic benefits that are referred to as benefits of concentration. Activities such as education, business and creative industries, which require interaction among a large number of individuals, specifically affect concentration.

Discussion and conclusion

As mentioned in the discussions presented in this chapter, many critics of smart growth often misconstrue the concept. Table 4.2 summarizes some of the most important criticisms of smart growth and possible responses to them.

In many cases, smart city critics select their units of measurement in such a way that can depict smart growth as unjustified but show a desirable picture of sprawl. For example, many methods, such as public transportation level of service (LOS), per capita annual congestion delay, average travel time, etc. are used to measure crowding and traffic congestion, some of which show mobility patterns and others the amount of accessibility. Dense regions have a higher LOS level but lower per capita delay because shorter distances and better travel options reduce the travel mileage by car. Though scattered areas have less traffic density, their per capita delay rates are higher due to the increase in travel mileage. The smart growth critics' claim that urban density increases traffic congestion is correct if calculations are done per square kilometre, but if we consider the per capita rate, that claim is

TABLE 4.2 Some of the most important criticisms of smart growth and the possible responses to them

The most important criticisms of smart growth	Responses
• Proponents sometimes exaggerate the benefits ofsSmart growth and the costs of sprawl. • Smart growth can have many unplanned implications. • There is uncertainty about the full costs of sprawl. • The idea of increasing public transport options in smart growth policies in order to reduce travel by car is a waste of time, because people deem travel by car the easiest way to travel.	• Support research to identify true benefits and costs and policies that reflect legitimate arguments. • A better understanding of the effects and the development of smart growth policies and initiatives is imperative. • Continue research and implement strategies that reflect market principles or help achieve strategic community goals. • Urban planners must investigate different options in order to find a way to improve the quality of footpaths and pedestrian traffic systems and other public transportation systems to a level where cars play a minor role in certain areas – especially in urban centres. • The solution for reducing crowding is not the extension of the transit network. Instead, roads should be closed to cars and more pedestrian and bicycle paths should be created. • In many cities, the public transport system needs to be upgraded for optimal performance. Such a scenario can lead to a clear reduction in energy consumption and emissions in car trips.
• Automobiles are the most efficient modes for many trips.	• Develop accessible communities and balanced transport systems that allow consumers to choose the best travel option for each type of trip. • Recognize that real efficiency accounts for all social impacts, not just from a single traveller's perspective. • Many smart growth strategies, such as increased access and shorter trips, can help save time. • Many critics do not consider smart growth's benefits, such as financial savings, increased physical activity of individuals, community cohesion and more environmental protection.
• Smart growth is a guideline for preventing urban growth and is always striving to prevent urban development.	• Smart growth should be seen as a concept that increases growth. The only thing that remains is the location, scale and method of such growth. • Smart growth strives to focus more on how to manage development, rather than limiting it. • Smart growth should be seen as a concept that seeks to achieve a balance between densification and market-based development in urban and suburban areas.

The most important criticisms of smart growth	*Responses*
• All developments must occur within existing urban areas, and the development or population growth in other areas is seen as sprawl.	• Smart growth principles can be applied in any rural, urban and suburban areas.
• Due to the rise of the suburbs in Europe, the stress on smart growth is in vain.	• Most European suburbs have more efficient transportation and use patterns than their American counterparts do, because they are more consistent with smart growth policies. In fact, smart growth can dramatically reduce land use and car travel; therefore, it can be beneficial even if land use and car travel increase in general.
• Regional population densities should be high and extensive, for example, 50,000 people per 2.6 square kilometres.	• Smart growth often deals with infill and cluster developments, not necessarily high density in vast areas.
• Traveling by car should be eliminated.	• Smart growth creates a more balanced and efficient transport system and, at the same time, matches the use of cars with most trips.
• A strict layer of rules and regulations aimed at slowing down the development process.	• If the regulations and conditions are corrected as intended by the concept of smart growth, the processes will be simpler. This way, beneficial projects get the necessary permissions faster, and their implementation becomes facilitated more easily.
• Smart growth supporters are seeking to modify construction guidelines and codes, zoning plans and uncontrollable regulations in order to simplify the methods and, in turn, make it easier to build projects reflecting smart growth characteristics.
• Regulatory strategies reduce the degree of consumers' choice and can have unplanned implications.	• To the extent possible, the application of smart growth strategies reduces regulations relying on market-based incentives and positive rewards such as increasing density, protecting green space and reducing travel by car.
• Increasing the development density by itself can increase traffic and local air pollution.	• A smart growth program should consider additional strategies along with increasing development density to improve accessibility, encourage optimal mobility and reduce car travels in the city.
• Many consumers value less dense housing in the suburbs and the car-oriented lifestyle.	• Consumers are given the choice to benefit from better land use and transport options.

(Continued)

TABLE 4.2 (Continued)

The most important criticisms of smart growth	Responses
• In terms of costs, transportation investments are not an effective way to reduce traffic and air pollution. • Strategies that reduce the amount of land available for development can increase housing costs. • People prefer to live in scattered and low-density contexts.	• Transportation will be more cost-effective if supported by other smart growth strategies that increase the operational efficiency of public transportation. • Execution of smart growth strategies increases the ability to pay for housing and transportation. • To understand the preferences of the people, it is necessary to consider other issues such as the time of daily trips, accessibility and the ability to live in local communities with an attractive and proper design.
• High density increases the possibility of an increase in social problems such as delinquency, poverty and insecurity.	• We should differentiate between density and congestion. • Smart growth can lead to a solution of social problems by increasing social interactions and economic opportunities for urban residents. • The use of smart growth policies can also help us use the social and economic benefits of concentration.
• The main goal of smart growth is to achieve the highest level of construction density.	• Smart growth pays special attention to infill and cluster developments. • According to the smart growth view, the existing densities increase in order to boost the capabilities of land but will not necessarily be considered at a maximum level.
• Smart growth leads to an increase in land ownership regulations and housing costs that, in turn, reduce housing affordability. • Implementation and effectiveness of smart growth policies are often long term.	• Reducing other regulations, including constraints on density, height, setback, use mixing and minimum parking requirements, can lead to a significant reduction in transportation costs and an increase in affordable housing. • Despite the long period needed for implementation and effectiveness of smart growth policies, these measures will improve the viability of local communities in the future and will have durable and long-lasting effects.

incorrect. Similarly, there are many possible ways to measure and compare smart growth effects on affordable housing, air pollution and health hazards. The way urban experts select and use them can provide different, and sometimes conflicting, findings.

On the other hand, urban scholars have developed a number of indicators in relation to smart growth that can reflect factors such as network compression, use mixing, connectivity and continuity of pathways and the diversity of transportation options. But many critics try to assess smart growth only based on the density of official and administrative units (such as the population of cities, suburbs or states) per area unit, an action that will lead to inaccurate results in most cases. Given the increase in the size of cities, there is a tendency towards an increase in population density that can lead to erroneous relationships and conclusions. More density data and comprehensive statistical analyses are needed to obtain meaningful information on smart growth effects. Smart growth does not require the conversion of a small town to a larger city, but sets proper requirements for each city or town with specific populations in order to increase network compression, use mixing, connectivity and continuity of pathways and the diversity of transportation options. Indeed, two regions can have the same population density, but one of them reflects smart growth while the other lacks it.

Many transport and land use factors are related to each other, so their simple analysis can lead to non-realistic conclusions. For example, density, congestion, travel distances, earnings and wages, travel differentiation, parking charges and rental rates all tend to increase with an increase in city size. But critics mistakenly state that smart growth exacerbates congestion delays, travel distances, operating costs for public transport and housing costs. Contrary to expectations, the likelihood of an increase in these costs will be greater in cases of sprawl and higher per capita annual car travel rates. Similarly, smart growth is more common in urban areas experiencing rapid population and economic growth, as well as high traffic congestion and housing costs. But that does not mean that it causes such problems. Smart growth can reduce many of these effects. However, critics often ignore these factors and assume that statistical correlations can be indicative of a cause-and-effect relationship.

The American Planning Association (2002) also points to some plausible mistakes that may occur if the urban management system focuses on growth management at a local scale. These mistakes can lead to inhibiting actions and ultimately have a negative impact on smart growth. A prominent example can be the ways in which the principles of smart growth interact with each other. As discussed earlier, the smart growth strategy follows a general and holistic approach to development issues. In other words, although each principle of smart growth can be used in isolation, the ideal result of this kind of development can only be realized when all the principles are seen together and in conjunction with each other, so that one principle does not put others at risk.

Only after considering the concept of smart growth, and in particular its principles, existing realities and status of smart growth designs around the world, can one express certain assumptions through which all the benefits advertised for smart growth can be achieved. Robert Liberty (2004), in an invaluable article consisting of his discussions with specialists on the state of smart growth today, addresses the successes and positive effects that smart growth can have on existing urban development and planning. Many experts interviewed by Liberty (2004) express positive views about smart growth's progress. These can be summarized as follows (Theart, 2007):

- Investments aimed at increasing the popularity of the smart growth approach can help achieve large-scale goals (Chen, 2004). Chen (2004) puts forward the hope that one day smart growth will be the main approach used in community-centred planning.
- In order to achieve smart growth principles, development policies at the local government level should be given special attention (Corbett, 2004). According to Corbett (2004), smart growth should become a routine matter in planning and development processes. Smart growth principles should be incorporated into spatial planning documents and continuously implemented and improved to create cumulative effects on urban sprawl.
- In the current context, many people in the United States continue to associate smart growth with "lack of growth" or "slow growth" (Harris, 2004). On the other hand, some criticize smart growth policies because they are not able to achieve the required results. According to Harris, the ideal for many people is still to achieve the American Dream of single-family houses in a relatively large area on a quiet countryside with two cars (and a garage) and enough food on the table. As time passes and new plans for smart development are presented, this dream has to be forgotten in order to create sustainable cities based on smart growth principles. Many people oppose the smart growth movement based on the sole reason that they still want to live on the basis of the American Dream without regard to its consequences and costs. Harris (2004) believes that the only way to promote the concept of smart growth is to listen to and understand the viewpoints of city administrators and community members who have been influenced by smart growth itself. This can help find ways to refine this approach and use the concept, respecting the wishes and experiences of all citizens.

The experiences arising out of the history of mankind suggest that any human model has weaknesses within itself. This chapter examined the critiques of smart growth based on such an understanding in order to provide a fair review of this approach and delineate challenges and obstacles to its implementation as much as possible. At the end, some responses to critiques of smart growth were provided. Although the authors are well aware of the potential shortcomings of this approach, it should be noted that smart growth today is a solid response to metropolitan issues, and it is possible to use it with a proper and logical implementation to achieve a far more favourable outcome with fewer drawbacks and weaknesses than is seen in the current conditions governing cities.

References

Abrams, R. F., & Ozdil, T. R. (2000). *Sharing the civic realm: Pedestrian adaptation in the post modern city, in urban lifestyles: Spaces-places-people.* Rotterdam: Brookfield.

American Planning Association. (2002). *Planning for smart growth 2002 state of states*. Chicago: American Planning Association.

Bashiri, L. (2011). *Determining urban residential density based on smart growth approach*. (Master of Arts dissertation), Tehran Art University.

Chen, D. (2004). What is the state of smart growth today, getting smart. *Newsletter of the Smart Growth Network*.

Corbett, J. (2004). What is the state of smart growth today, getting smart. *Newsletter of the Smart Growth Network*.

Cox, W. (2011). The housing crash and smart growth. *National Center for Policy and Analysis, 335*, 1–16.

Cox, W., & Utt, J. (2004). *The costs of sprawl reconsidered: What the data really show*. Washington, DC: Backgrounder.

Gordon, P., & Richardson, H. (2000). *Critiquing sprawl's critics*. Retrieved from Washington: https://object.cato.org/pubs/pas/pa365.pdf

Gordon, P., & Richardson, H. W. (2001). *The geography of transportation and land use*. Paper presented at the Smarter Growth: Market-Based Strategies for Land-Use Planning in the 21st, Holcombe.

Harris, M. (2004). What is the state of smart growth today, getting smart. *Newsletter of the Smart Growth Network*.

HosseinZadeh Delir, K, & Safari, F. (2012). The effect of smart planning on urban spatial planning. *Journal of Geography and Urban Development*, (1) (Spring and Summer).

Liberty, R. (2004). What is the state of smart growth today, getting smartwhat is the state of smart growth today, getting smart. *Newsletter of the Smart Growth Network*.

Litman, T. (2011). *Evaluating criticism of smart growth*. Retrieved from Canada: https://www.vtpi.org/sgcritics.pdf.

Litman, T. (2012). *Smart congestion relief: Comprehensive analysis of traffic congestion costs and congestion reduction benefits*. Paper presented at the Transportation Research Board Annual Meeting. www.vtpi.org/cong_relief.pdf

Litman, T. (2015). *Evaluating criticism of smart growth*. Washington, DC: Victoria Transport Policy Institute.

O'Niell, D. (1999). *Smart growth, myth and fact*. Retrieved from Washington DC.

O'Tool, R. (2001). The folly of smart growth, Thoreau Institute, Regulations. Retrieved from www.cato.org

Rahnama, M. R., & Abbas Zadeh, G. (2006). Comparative study of distribution/compression grading in metropolitan Sydney and Mashhad. *Geography and Regional Development Magazine*.

Sheehan, M. O. M. (2001). *City limits: Putting the breaks on sprawl*. Retrieved from Washington: http://www.worldwatch.org/system/files/WP156.pdf

Theart, A. (2007). *Smart growth: A sustainable solution for our cities?* (Master of Philosophy in Sustainable Development Planning), University of Stellenbosch, South Africa.

Thompson, J. J. (2012). *Growing with the flow: planning for smart growth in ontario through water & wastewater infrastructure service provision*. (Master of Planning), Ryerson University, Toronto.

USEPA. (2011). *Essential smart growth fixes for urban and suburban zoning codes*. Retrieved from Washington, DC, www.epa.gov/smartgrowth

Yang, F. (2009). *If 'Smart' is 'Sustainable'? an analysis of smart growth policies and its successful practices*. (Master of Community and Regional Planning), Iowa State University, Iowa.

Ziari, K. A., Hatami Nejad, H., & Turkmennya, N. (2012). Introduction on urban smart growth theory. *Shahrdariha Journal, 12*(104).

5

A COMPREHENSIVE CHECKLIST OF GENERALIZABLE AND ACHIEVABLE GOALS, STRATEGIES AND POLICIES FOR SMART GROWTH

(With an emphasis on pedestrian-oriented transportation)

Introduction

As mentioned in previous chapters, this book seeks to find a logical framework for building sustainable urban spaces based on smart growth, with the goal of improving the quality of urban transport in a sustainable and people-oriented way. Accordingly, the first chapter examined the theoretical approaches and applied concepts related to smart growth (such as definitions, history, formation stages, features and benefits, as well as tools, strategies and techniques for smart growth). In the second chapter, the smart growth approach was compared to urban sprawl in different use cases. Subsequently, the third chapter focused on global experiences in the evaluation of smart growth–based urban development plans and policies. Finally, the fourth chapter presented a review of the criticisms of this approach.

In the present chapter, reasonable and achievable goals, strategies and policies in this area will be devised using the materials presented in the last four chapters, as well as extracting the most important operational texts in the field of smart transportation.

Reducing reliance on personal transportation and providing a variety of transportation options

One of the most basic goals of smart growth in relation to an urban transport system is that of reducing reliance on personal transportation and providing a variety of transportation options (Figure 5.1). This goal can reduce the consumption of fossil fuels and, consequently, their pollutants, such as carbon derivatives and greenhouse gases. Extensive research has also been done on changing the mode of personal transportation to other forms, such as walking and public transport, and their impact on the health of citizens. Providing various transportation options as one of the key goals of smart city development amounts to providing

FIGURE 5.1 Various transportation options in New York City

Source: Moreira, 2009

citizens with more access to housing, neighbourhoods, stores and transportation systems. Cities are expected to satisfy these needs through a variety of transportation options.

According to traffic studies, the capacities of new roads are filled much faster than their construction speed. In other words, increasing demand in the area of transportation indicates that, as new large roads are built, people increase the use of cars to take advantage of the new facilities for transportation. Some studies show that only 60 to 90 percent of the capacity of new roads is used within five years after they are opened (Chen, Yang, Lo, & Tang, 2002)

Research shows that though the traffic behaviour and tendency of people can change in the short term from public transportation and group (shared) trips to travel through new roads, in the long run, the desire to use private cars will be reduced by providing better access to networks and different transportation options to various metropolitan areas, especially the suburbs (Chen et al., 2002).

In response to this issue, cities have implemented new strategies for transportation planning. These strategies include increasing compatibility between land-use and public transport system, increasing access to high-quality transportation services, increasing the number of transportation vehicles, linking transportation networks and establishing a reasonable link between pedestrians, cars, transportation, and road facilities. In short, new transportation planning strategies create a hybrid model with supporting land use patterns that is able to create a variety of transportation options (Curtis & James, 2004). In general, the quantitative and qualitative promotion of public transport is considered as a very important strategy for achieving sustainable transport under the concept of smart urban growth.

Implementation of this strategy through the application of quantitative policies such as "increasing the number of active buses in the urban bus system, increasing the number of subway lines," and policies for increasing the quality of public transport such as suitable climatic design for bus and subway stations, informing people of the schedules of buses and subway trains, and presenting online travel information for travellers so that they can better plan for the use of the public transport network (Figures 5.2 and 5.3).

FIGURE 5.2   Utilizing convenient bus stops in terms of climatic comfort in Curitiba, Brazil, can be effective in increasing the attractiveness and use of the public transport system

Source: Aickin, 2008

FIGURE 5.3 Creating facilities such as small urban libraries in Paris can be effective in increasing the attractiveness and use of the public transport system

Source: Maclennan, 2019

Other policies that can be used for the quantitative and qualitative promotion of the public transport system include improving the quality of public transport and equipping them with heating, cooling and ventilation equipment, using natural gas vehicles (NGVs) in the public transport system, and equipping public transport stations with solar energy systems.

Relying more on the bike system is a strategy that can lead to maximum use by citizens, if accompanied by feasible policies such as increased connectivity and combination of bike system with other transportation modes (Figure 5.4).

In order to combine the bike system with other types of public transport, it is necessary to create opportunities for urban bike riders to carry their bikes into subway stations or even trains, or to connect bikes to transit buses to be carried to different places (Figure 5.5 and 5.6).

Another policy that can help to further realize this strategy is the allocation of bike lanes with a suitable width between pedestrian paths and streets. It should be noted that the mere allocation of such lanes to bikes is not enough, and they must also be secured in various ways for the safety and security of their users. Therefore, the use of shields or buffers along the bike lanes can prevent motor vehicles from entering them and provide an optimum level of safety for their users during busy hours and times of traffic congestion (Figure 5.7). Also, the

FIGURE 5.4 An example of the connectivity of different public transport systems to one another

Source: Morgan, 2013

FIGURE 5.5 AND 5.6 The possibility to move bikes by other public transport systems to encourage more people to use bikes

Source: Viriyincy, 2006, 2011

FIGURE 5.5 AND 5.6 (Continued)

FIGURE 5.7 Creating special bike paths with secure buffers will increase the desire to use them

Source: Krueger, 2011

use of smart traffic light control systems will be to the benefit of pedestrians and cyclists in the short run and is a good incentive to encourage citizens to use the bike system.

The next policy that can be used to achieve this strategy is smart design of the general shape and form of bike lanes. For example, in designing bike paths, it is necessary to avoid circuitous paths as much as possible (especially in the paths leading to intersections) and seek to design direct bike paths. The shortest routes for the bike system should be used to link the main functional nodes of the city or the main centres of travel production.

In general, it seems that the use of bikes in comparison to the use of personal vehicles not only increases safety and promotes physical health but also is cheap and provides the shortest possible route and, hence, the shortest travel time that can add to the attractiveness of the use of this transport system.

The design of bike paths should also be of sufficient visual appeal. Prioritizing visual attractiveness can include natural scenery like the margins of parks, valleys, canals, etc. and if such a landscape does not exist, a green strip of climate-compatible plants or trees on the bike lane edges should be designed. In addition to creating visual beauty, it can act as a buffer between pedestrian and car traffic and the bike paths.

Table 5.1 illustrates summary of all goals, strategies and policies described in this section.

TABLE 5.1 Summary of goals, strategies and policies aimed at reducing reliance on personal transportation and providing various transportation options

Objectives	Strategies	Policies
– Reducing reliance on personal transportation – Providing a variety of transportation options – Reducing the consumption of fossil fuels and, consequently, their pollutants, such as carbon derivatives and greenhouse gases – Promoting people's health	– Increasing compatibility between land use and public transport system – Increasing the number of transportation vehicles – Increasing access to high-quality transportation services – The quantitative and qualitative promotion of public transport	– Increasing the number of active buses in the urban bus system – Increasing the number of subway lines – Suitable climatic design for bus and subway stations – Informing people of the schedules of buses and subway trains – Presenting online travel information for travellers so that they can better plan for the use of the public transport network – Improving the quality of public transport and equipping them with heating, cooling and ventilation equipment – Using natural gas vehicles (NGVs) in the public transport system – Equipping public transport stations with solar energy systems

Objectives	Strategies	Policies
	− Creating and developing bike systems	− Combining the bike system with other types of public transport − Allocating bike lanes with a suitable width between pedestrian paths and streets − Maintaining and enhancing the safety and security level for users of this transport system using measures such as a shield or buffer along bike paths − Using smart light control systems at intersections − Using smart designs for the shape and form of bike paths and avoiding designing circuitous paths as much as possible (especially in the paths leading to intersections) − Using the shortest routes for the bike system to link the main functional nodes of the city or the main centres of travel production − Creating attractive and pleasant bike paths by prioritizing the use of natural landscapes, such as the margins of urban parks, valleys, canals, etc. − Using green plant strips and planting climate-compatible plants or trees on the bike lane edges

Optimal land use in line with targeted development in existing urban areas

Urban managers are responsible for planning and investing in the intra-city transport sector, and they can play an important role in selecting the transportation types used (e.g. car-oriented transportation systems, bikes, motorcycles or pedestrians paths) and in the targeted allocation of facilities in this area.

But the important point is that successful and efficient design, planning and investment in the field of intra-city transportation are contingent upon the proper selection and deployment of land uses. For example, the pedestrian network would be an inefficient option if it is not close to the intended destinations of the target group. Therefore, a smart investigation of the proposals made for creating new uses, or changing the function of existing uses, and the proper location of uses in the vicinity of current walkways, alongside designing the pedestrian network based on existing uses in urban areas, is essential.

On the other hand, providing transportation services in low-density developments is costly and expensive, because workplace and shopping areas are far from each other in these areas, leading to an increase in the volume of traffic flow (Figures 5.8).

Urban managers can enhance the efficiency of investment in transportation by ensuring the compatibility of the transportation system with development (Ferguson, 1990). In this regard, priority should be given to projects that properly provide for mixing land uses in their design. Governments can also support cross-sectoral investment plans for areas involving mixed land uses by providing incentives and grants to local administrations. They can also provide incentives for neighbourhoods that have accumulated key uses in walkable places or those for neighbourhoods that have safe passageways for schoolchildren.

Cities can also benefit from flexibility in the rules that encourage the participation of the private and public sector in transportation and development investments. In this regard, city managers, for example, can help with the redevelopment of brown fields and barren lands and the exploitation of existing transportation networks through the use of infill and intensive development policies.

The summary of goals, strategies and policies for this section can be seen in Table 5.2.

FIGURE 5.8 Delivering transportation services in low-density urban development is often extremely costly and economically infeasible

Source: Jouan, 2009

TABLE 5.2 Summary of goals, strategies and policies focusing on the optimal use of land for targeted development in existing urban areas

Objectives	Strategies	Policies
Optimal use of land for targeted development in existing urban areas	Compatibility of the transportation system and urban development	– Implementing intensive and infill development policies – Redevelopment of brown fields and barren lands in order to take advantage of the capacity of existing transportation networks – Prioritizing projects that properly provide for mixing land uses in their design
	Strengthening local transport systems to adequately respond to the daily needs of user groups	– Providing incentives and grants to local administrations to support cross-sectoral investments in areas involving mixed uses – Providing incentives for urban areas with safe passageways for schoolchildren – Providing incentives for neighbourhoods that have attracted crowds and key uses in walkable places
	Adaptation of use location with existing and proposed transportation systems	– Locating new uses or changing existing uses in the vicinity of existing transportation networks – Choosing the type and routes of transportation networks based on the performance of existing uses in urban areas – Functional performance overlapping of various urban areas and full adaptation of the transport network to existing areas – Mixed urban uses
	Paying attention to the optimal density of the urban areas	– Optimal loading of building density in order to maximize the use of land in urban areas, thereby reducing the need for new trans-shipping networks – Paying more attention to the quality and diversity of urban transport networks, rather than to the quantity of areas assigned to them

Improvement and modification of level of service standards in areas with public transport

The quality of the level of service (LOS) is a quantitative scale of service measuring a street or intersection's service-providing level. LOS is a ratio of time to flow rate of traffic. However, its characteristics are not limited to the number of passing vehicles in a duration of time in a specific path; rather, the freedom to act and change the direction of movement (movement manoeuvre) also plays a role in defining its various levels (Papacostas & Prevedouros, 1993) An important point is that in some cases, when the traffic flow rate increases, municipalities plan to renovate the street based on a cliché (as the most simple solution to handle high traffic congestion through increasing the width of street) in order to improve the traffic situation. However, sometimes these solutions (e.g., widening the street) will not necessarily solve the problem of congestion. At many times, lively and flourishing parts of the city have a high traffic flow rate (Newman & Kenworthy, 2006). This is a sign of vitality, especially in the central parts of the city. Similarly, light traffic can be a sign of the gradual decline and destruction of the central business district (CBD). Additionally, strict compliance with service-level standards in per capita form, particularly in areas that lack high-quality public transport, is not recommended. In these areas, the reliance on service-level standards does not reflect access to public transport services (Dowling et al., 2008).

As discussed earlier, though the matching of street service level with traffic flow can entail an efficient urban flow, it cannot explain the quality of urban transport alone. If we assume that the area benefits from a high-quality public transport system, then a strategic plan for achieving the appropriate service level will be necessary. "Creating a balance between the service level and traffic flow" is a strategy that matters in the context of smart growth. This balance will not necessarily indicate the need to widen the street, and in fact considers both the dynamics resulting from the distribution of population density and the traffic flow through the street.

A summary of goals, strategies and policies for improving and modifying road service-level standards can be seen in Table 5.3

TABLE 5.3 Summary of goals, strategies and policies with a focus on improving and modifying road service-level standards in public transport areas

Objectives	Strategies	Policies
Improving and modifying road service-level standards in public transport areas	Creating a balance between the road's service level and traffic flow	– Improving the level of service based on a local community's needs resulting from their activity and residence – Improving the bike level of service based on bike ownership in the area – Improving the personal car level of service based on the average daily traffic of personal cars
	Upgrading the pathway to the level of service of a full street	– Providing a bike route – Providing a public transit lane – Providing a route for rail transport at zero level (ground level) – Providing sidewalks on the edges of the streets – Improving economic vitality and prosperity in urban centres

Proper connection of the local street network to higher transportation levels

Hierarchical street networks consist of local, collector and distributor, arterial and freeways. Despite the merits of this type of street design, traffic flow only remains on a small number of freeways in the case of accidents or traffic jams. On the other hand, focusing traffic on the main roads does not provide a good environment for pedestrians, and at a higher level, for residential development (Jiang & Claramunt, 2004.) Consequently, the space among these urban corridors (arterial and freeways) will be accessible only by car.

The strategy for creating a *fused grid street layout* (Figure 5.9) out of small-scale streets can lead to a high amount of traffic movement and can help vehicles avoid long-term delays caused by U-turns on large crossroads. In such a layout, auxiliary

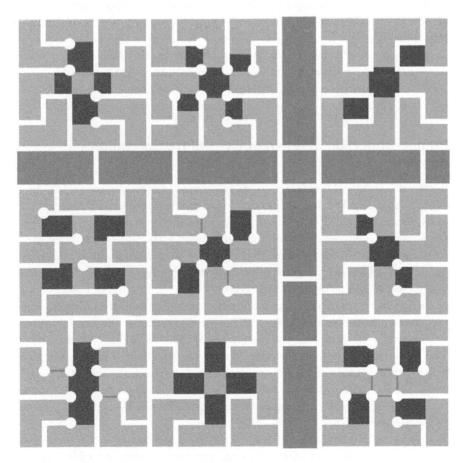

FIGURE 5.9   An example of road design in four neighbourhoods based on the fused grid street layout

Source: Gomes Franco, 2007

TABLE 5.4 Summary of goals, strategies and policies focusing on the correct connection of local area networks to higher transportation levels

Objectives	Strategies	Policies
Connecting local area networks (with two to four lines) to higher transportation levels	Creating a fused grid street layout out of small-scale streets	– Connecting the main streets of neighbourhoods to one another and to a higher level of transportation – Maintaining the internal tranquillity of neighbourhoods using impasses – Maintaining the role and size of auxiliary streets proportional to the neighbourhood level

streets create short-cut spaces among different points of collector and distributor streets and can free up a major part of the main road. This opportunity can facilitate the flow of the main route in favour of buses for public transport. It should be noted, however, that by creating a wave network, auxiliary streets become active during times of traffic, but the role and size of these streets continue to be proportional to the neighbourhood level (Grammenos & Pidgeon, 2005).

Table 5.4 describes policies for connecting local area network to higher transportation levels.

Proper connection of different types of transportation networks

Before the last two decades, most transportation networks (or systems) were designed in a detached and unconnected manner. Establishing sufficient connections among different types of transportation modes is an important foundation for achieving multi-modal transportation systems. It should be noted that every trip begins and ends with walking. In cases where pedestrian traffic does not have good access to the starting point of the trip, an inappropriate level of transportation is created, and accessibility and diversity diminish. All types of transportation in the city are considered appropriate when they are connected to each other. For example, bike places at public transport stations increase the number of passengers and users of public transport services (Saelens, Sallis, & Frank, 2003). The transportation systems that carry their own bikes have a higher efficiency because they extend the breadth of the trip's origin and destination. Similarly, when destinations are closer and there are sidewalks between different locations, car-based trips will be made more efficient and effective by connecting to the pedestrian network.

One of the important strategies for achieving smart transportation is to establish a connection between different public transport systems (see Table 5.5). For example, citizens should be able to move easily between the transit buses and subway system. Because the development of subway lines is the costliest compared to

TABLE 5.5 Summary of goals, strategies and policies with a focus on the right connections of different types of transportation networks to one another

Objectives	Strategies	Policies
Connecting different types of transportation networks to each other	Establishing connections among different public transportation systems	– Creating bike places at public transport stations – Creating sidewalks among different locations to enhance the efficiency of trips – Facilitating citizens' trips among bus and subway systems
	Making public transit stations closer to sources of urban trips	– Expanding public transit stations at short distances (up to a maximum of 300 metres) from one another – Increasing residential density around the stations – Locating public transit stations within 400 metres of the residential area

other transportation types, the construction and establishment of subway stations and subway lines for all urban neighbourhoods are infeasible and even sometimes impossible given various issues such as soil quality, the possibility of excavation, tunnel-boring issues, etc. As a result, for those citizens whose destinations do not have a subway station, the previously mentioned connections can be used to get to the nearest subway station and enter the transit bus system and continue their trip with the most desirable mode of public transport. To achieve this strategy, we should expand public transit stations at short distances (up to a maximum of 300 metres) so that citizens can easily choose their own system from among different public transport systems (CLADCP, 2008). In determining the scope of this policy, we should aim at increasing the options for citizens and consider the visibility and connection among public transport systems. This means that the placement of public transport stations at optimal distances will help persons who are unfamiliar with the environment to easily reach the subway or bus station and change the mode of public transport with a few minutes of walking.

In addition to close links between different types of urban transportation, public transit stations should be close to the sources of citizens' urban trips. This provides them with easy access to public transport and thereby increases the efficiency of the system. This can be achieved through the increase in residential density (as one of the main starting points of urban trips) around the stations, as well as the location of public transport stations within a radius of 400 meters from the residential area that place points of access to public transport at desirable distances to destinations. This way, more people will be able to easily access the system (WRCG, 2012).

Creating an area for centres of activity around transportation systems

Achieving an efficient public transportation service requires proper design and planning for routes, qualitative and quantitative promotion of public transport pillars, etc. One of the measures that can contribute to the efficiency of this system is supporting the public transport service through land use. Establishing a higher density residential development around transportation stations (Figure 5.10) can provide access to the public transport system for more households (Burden et al., 2009).

In addition, if services and other facilities are located near transportation stations, the public transport would be more efficient and attract more people. Such services and facilities include local services like childcare services, as well as facilities for daily trips, such as laundry and ironing facilities and convenience stores. This way, citizens can do other everyday activities, such as shopping, while using public transport. Table 5.6 illustrates two policies regarding the goal of supporting public transportation through land use planning.

FIGURE 5.10 Placing a train station near a high-density residential area, Dubai, United Arab Emirates

Source: Mogliani, 2019

TABLE 5.6 Summary of objectives, strategies and policies focusing on creating an area for activity centres

Objectives	Strategies	Policies
Supporting public transportation services through land use	Creating an area or zone for activity centres around the transportation systems	– Developing residential buildings with higher densities around transportation stations – Locating services like childcare services, as well as facilities for daily trips, such as laundry and ironing facilities, and convenience stores along with transportation stations in order to increase the efficiency and attractiveness of public transport for citizens

TABLE 5.7 Summary of goals, strategies and policies focusing on creating pleasant and attractive walkways

Objectives	Strategies	Policies
Creating pleasant and attractive walkways in all new developments in the city	Mixing uses, small blocks and close destinations	– A balanced mix of transportation options – The requirement for construction of sidewalks in new buildings by the government

Making pleasant and attractive walkways

Sidewalks are essential for creating a safe and unobtrusive walking environment and therefore a balanced mix of transport options (Burden et al., 2009). However, in the first half of the twentieth century, with the increasing dependence on cars, many new streets were built without sidewalks. Sidewalks alone cannot attract pedestrians. In other words, to turn walking into leisure-recreational practice (see Table 5.7), the use of other elements such as mixed uses, small blocks and close destinations is required. Local governments should ensure that new construction includes standard and quality sidewalks so that citizens can use them for walking or riding bikes.

Satisfying parking needs

Creating parking is costly, and location plays an important role in creating a balanced transport system. Optimizing access to parking for local destinations affects the individual selection of driving, walking or using public transport. Parking also

TABLE 5.8 Summary of goals, strategies and policies focusing on meeting parking needs

Objectives	Strategies	Policies
Meeting parking needs	Optimize access to parking in local destinations	– Using street parking to provide the required stopping distance by municipalities – Reducing the number of parking spaces required for new construction in areas of public transport or mixed-use areas – Specifying priority areas for parking – Reducing the cost of parking for shared cars – Allowing non-cash payment of parking fees for employees – Allocating income resulting from parking taxes to workers – Encouraging builders to construct parking behind the building or inside the yard

influences the financial profitability and form of development and the provision of a "multi-modal transport system". For example, in a car-oriented infill development project, the construction of multiple car parks will increase the development costs and reduce the ability of the project to support public transport.

Municipalities can affect people's decision to select the type of transportation system through parking. Municipalities may use on-street parking or reduce the number of parking lots needed for new construction in areas of public transport or mixed-use areas. They can also take other effective measures such as allocating priority areas for parking, reducing the cost of parking for shared cars, allowing non-cash payment of parking charges for employees or allocating the income resulting from taxes on parking fees to workers (Burden et al., 2009). They can encourage builders to construct parking behind the building or in the yard. Such parking performs much better than parking in front of buildings or surface parking and prevents the disorder of public spaces during busy hours. Seven policies regarding to meeting parking needs described in Table 5.8.

Granting incentives to reduce traffic during busy hours

The patterns of daily trips are largely influenced by a series of individual decisions for moving from residential to activity areas. For example, the time and means by which a person travels from his place of residence to work affect the dominant pattern of daily trips. Employers in both the public and private sectors are affected by these decisions. In the past decade, employers, unlike local government and municipalities, have been working to achieve more efficient management of their

TABLE 5.9 Summary of goals, strategies and policies focusing on giving incentives to reduce traffic at peak hours

Objectives	Strategies	Policies
Giving incentives to reduce traffic at peak hours	Giving information and incentives for commuting options	– Integrating different departments and offices at locations close to subway stations – Providing housing in partnership with employers for employees who want to live close to their workplace – Providing the opportunity to work at home through teleworking – Providing flexible working hours through flexible times

employees' daily commuting, and they are doing so for robust business reasons. For example, in Atlanta, BellSouth's concerns about the commuting problems of its employees led to the merger of more than 70 offices and collecting them in three locations close to subway stations.

Municipalities can have an important role to play in increasing such efforts by providing the necessary information and giving incentives for employer-supported transport options. Granting subsidies for the use of public transport and housing, in partnership with employers, for employees who are close to their workplace can help reduce traffic during busy hours (CLADCP, 2008). In addition, among the various transportation options, one option is for employees to remain at home. Employers who give employees the opportunity to work at home through telecommuting (telework) or who create opportunities for intensive working hours through flexible times (also called flex-time, this refers to a change in work hours that allows employees to go to or exit from work hours at times other than peak hours) can affect the selection of optimal and appropriate transportation options (Table 5.9).

Coordinating different transportation services to take advantage of the full benefits of neighbourhoods and developments supporting public transport

Making land use compatible with transportation

Land use that is compatible with transportation paves the way for future development and new transportation services. There can also be good opportunities within the existing public transport system. Municipalities can assess the existing public transport system and create bus stations or bus routes with good access to high-quality walking paths in high-density and mixed-use areas. Local governments can, in addition to creating stations in areas with supporting land use, set transportation goals based on demographic and economic factors (Handy, 2005). For example, neighbourhoods with a higher percentage of young people and students may need more cycling, and strategic decisions can help improve regional impacts and create a balance between housing and employment beyond the region (Morris, 1997). For full success, the nature of work trips and other trips should be taken into consideration when deciding on time and cost plans.

In the preceding section, the strategic importance of the bike system and the public transport system was mentioned, but the form of a sustainable transport system is completed with sidewalks. Promoting the role of sidewalks among transportation modes is considered a separate goal here because of its importance. This goal cannot be achieved by the mere allocation of sidewalks along the streets. Encouraging citizens to choose walking for daily trips requires the adoption of a set of specific goals, strategies and policies in the urban planning decision tree.

The use of a wide range of quality housing options for all income groups

The study of large cities and small towns alike suggests that a combination of different housing options (such as apartments, single-family housing units, etc.) can be found in urban centres, and as we move away from the city centres, we will encounter large areas that merely offer houses. This creates a situation where only a certain type of residence is formed in each neighbourhood and leads to houses that many people cannot afford to buy or to housing types that, despite their proximity to workplace, school or other distinct activities, cannot adequately respond to the needs of many households (Theart, 2007). Providing a wide range of housing options for citizens will make households with different financial abilities choose the options appropriate for them. In this way, such households are more committed to maintaining their local area, and the sense of a connected community (with ideal social unity) is achieved (U.S. Environmental Protection Agency, 2002). Providing a wide range of housing options to all households allows them to find their place in the smart growth community (e.g. a garden, apartment, traditional houses on the outskirts, etc.) and thus adapt themselves to the present-day conditions (Rahnama & Abbas Zadeh, 2006). A strategy for achieving this goal is providing high-quality housing for people at all income levels. Realization of this goal depends on various policies, including modification of the structure of single-family housing and the expansion of housing options, including flats, residential complexes, and multi-family housing. Also, combining single-family and multi-family residences in existing locations can reduce poverty, increase optimum density and effective location in relation to residence and work and lead to better satisfaction of the residents' needs, equity in access to the transportation network,[1] promotion of public transport and many other benefits to the heterogeneous population of neighbourhoods. In compact and dense contexts, offering different housing options of varying sizes and prices (in the form of duplex townhouses, flats and separate single-family detached housing) within walking distance of shops, offices, recreational areas, as well as the public transport system, help us ensure that a wider range of individuals and groups in the community are able to live in the residential area (Theart, 2007, p. 9).

The policy of providing affordable housing for low-income households can be considered an appropriate response to demand in the housing market or urban facilities and installations market, and can increase the city's economic efficiency and prevent the formation of poor housing such as slums (Litman, 2005). Another policy that can be taken into consideration here is tailor-made housing for the elderly and the disabled people. Table 5.10 illustrates summarised goals, strategies and policies for establishing coordination among different transport services.

TABLE 5.10 Summary of goals, strategies and policies focused on establishing coordination among different transport services

Objectives	Strategies	Policies
Establishing coordination among different transport services	Creating land use compatible with public transport	– Placing bus stations or routes in high-density areas – Locating stations in areas with supporting land use – Locating bus stations in mixed-use areas with access to high-quality walking paths – Setting transport system goals based on population and economic factors of the target community – Considering the nature of trips and travel patterns and making decisions accordingly
	Providing a wide range of high-quality housing options for all income groups	– Changing the structure of single-family housing and expanding housing options, including flats, residential complexes and multi-family housing – Combining single-family and multi-family housing in existing neighbourhoods – Providing different housing options in various sizes and prices (in the form of duplexes, flats and single-family detached housing) in compact and dense contexts – Providing affordable housing for low-income households – Providing homes suitable for the elderly and disabled people

Creating walkable communities

Before the 1900s, cities used to focus on walking and were designed in such a way that people could walk to their destinations. However, in the last 50 years, dispersed development patterns and the separation of uses have led to greater reliance on personal cars and the elimination of many features of walkable communities. Today, many traffic and road engineers propose the possibility of leaving one side of the street without a sidewalk. Many engineers and builders believe that creating sidewalks does not have much economic value when no one uses them. However, inclusion of good features and qualities in the sidewalks, the proper mix of uses and densities, compact street intersections and neighbourhoods convenient for citizens all affect the walkability of cities (Morris, 2013).

Walkable communities are the best model for achieving smart growth goals, because they increase mobility, reduce negative environmental impacts, increase economic growth and support social integration in neighbourhoods. Cities can be designed to make walking to destinations the most suitable option, and this will improve access to services for one-third of the population that is too young, too old or too poor to drive (Handy, 2005). Cities that seek to increase walking access have many environmental benefits. For example, the quality of air will increase as people reduce the use of cars to reach their destinations and walk instead. In addition, measures such as the design of narrower streets, the creation of on-street parking instead of off-street ones and the creation of sidewalks in parking lots reduce impermeable levels and decrease surface flow and runoff.

These strategies have many economic benefits, as cities with better air quality and healthier water impose fewer public charges on taxpayers. Other social and economic benefits associated with walkable cities include low transport costs, increased personal health and increased user selectivity. When all these benefits of walkable cities are recognized, more cities will force the public and private sectors to develop and construct walkable neighbourhoods (Figure 5.11).

Appropriate mix of uses and density

Many public and private sectors prevent the construction of walkable communities. Some of the common land use laws make it difficult to mix uses, thus increasing the length of trips and turning walking into an inappropriate and inefficient option. These land use policies, supported by some private-sector financial policies, consider mixed-use development more dangerous than single-use development. Many cities, especially cities that are large in scope and dependent on cars, support projects that reduce walking. Many of the current designs offer wide streets, which have large-scale blocks, less walkability and limited infrastructure for sidewalks (Handy, 2005). Also, the common designs of residential construction work as a major obstacle to walking. Setback criteria, criteria for large plots, indirect streets and impasses increase the distance between sidewalks and destinations. There are similar hurdles within commercial designs (Evans, 2002). Many administrative

FIGURE 5.11   An example of a walkable neighbourhood

Source: La Citta Vita, 2011

buildings, retail stores, hotels and other commercial facilities are surrounded by large parking lots that force pedestrians to cross between parked and moving cars.

These obstacles indicate that land use and design play a key role in encouraging walking. Cities can increase the quantity and quality of walkability by creating multiple destinations and placing uses in close proximity to each other. It should be noted that sidewalks and walking create an equilibrium among all forms of transportation. This adjusts the scale and size of buildings and corridors and brings them in proportion with the human scale. The following is a list of strategies that can help cities achieve this basic goal.

Zoning based on form

Configuration building forms is an important measure that can be used as a substitute for conventional zoning (based on use). In these regulations, large buildings are placed among other large buildings, medium-size buildings are placed among other medium-size buildings and, similarly, small buildings among small buildings (Breheny, 1992). Generally, as we move outward from the centre of the neighbourhoods, buildings become shorter and the percentage of land occupancy decreases. These different forms usually result in different uses, without turning this phenomenon into a law. Together, the various types of buildings in a neighbourhood prevent monocultural physical and social forms and reduce the similarity that ultimately results in destruction. Similar setbacks and suitable parking places can, in spite of the potential diversity in the mixing of uses, lead to a kind of coordination. The streets, with the exception of rare cases, should be symmetrical and have similar types of buildings on either side of them, and the change in zoning should take place in the middle of the blocks along the back of the land lots that is not visible. Thus, a homogeneous street view can be achieved with incompatible blocks (Breheny, 1992).

Retrofit existing streets and sidewalks to create walkable neighbourhoods

Retrofitting existing streets and sidewalks and adding facilities for pedestrians need to be budgeted for, just as other development activities do. In today's economic atmosphere, where cities sometimes remove sidewalks, pedestrian crossings and other factors that encourage walking, for many, the importance of the funds spent on these changes has not yet been properly understood. State governments can play an important role in directing financial resources and technical support to help these local efforts. Targeted use of government and municipal funding in the transportation sector can help communities implement initial retrofitting projects, which can reveal the benefits of walking improvement and lead to more local financial support for pedestrian-oriented neighbourhoods (Brambilla & Longo, 1977). For example, the state of Illinois began a program called "Illinois Tomorrow: Balanced Growth for a Better Quality of Life", which granted about $3.7 million to local communities to promote land use, transportation and infrastructure programs (Illinois Agency for Planning, 2010). Most of these state grants are used to improve walkability on local streets through planning, pedestrian-oriented restoration and public transport–based development. Federal funds like those related to the Transportation Equity Act for the 21st Century support the combination of cycling and walking as the main means of transportation (FHWA, 1998). Most importantly, this can improve the ability of cities to invest in projects that encourage safety and the possibility of cycling and walking on daily trips. Cycling and walking projects are eligible to receive funding from all federal transportation programs, including those on highway construction and increased security. In the years before the adoption of the Transport Equity Act, the federal government's spending on walking and cycling facilities was about $4 to $6 million a year. Since then, the amount of federal funding has increased significantly (more than $296 million in 2000) (Rodriguez & Goerman, 2004).

Improving the comfort, convenience and safety of sidewalks

Improving the comfort, convenience and safety of sidewalks requires breaking down this strategy into a number of logical, feasible and implementable policies. In this regard, five policies are discussed:

a) **Creating sidewalks with the proper width.** The proportionality of the size of the sidewalks to the population passing through them is of great importance.

Accordingly, if the width of the path exceeds the pedestrian flow demand, it leads to insecurity in the late hours of the day and imposes costs for maintaining an unused space. Also, if the path is too narrow, the traffic volume exceeds its capacity, and the resulting congestion decreases the quality of the surrounding urban space. Also, maintaining the quality and cleanliness of the path will face problems, and any extra repair or cleaning done to respond to the high pedestrian flow rate will eliminate the feeling of comfort and pleasure of walking for users. Therefore, determining an optimal level that is cost-effective and can respond to users' needs will be of great importance in achieving a smart urban sidewalk strategy. According to studies by the American Planning Association, "a width of at least 1.5 meters for sidewalks" is essential for the design of urban roads (CLADCP, 2008).

b) **Designing sidewalks in a coherent and continuous manner (as straight as possible).** The Los Angeles Department of City Planning, in its pedestrian volume of *Principles of Design for Smart Growth*, states that based on Newton's first rule we should avoid changing the sidewalk direction to the greatest extent possible (Figure 5.12). This organization believes that direct and continuous routes will attract more citizens to walk in their everyday trips (CLADCP, 2008).

c) **Establishing a buffer between the sidewalk and the vehicle path.** Creating a buffer between the sidewalk and the vehicle path using the natural landscape or street features can increase the perceived safety of pedestrians. Research conducted by Niehoff Studio in 2010 shows that pedestrians' use of sidewalks increases if there are urban features, metal benches, postal kiosks, telephones or natural elements like trees on the border between the sidewalk and the street (Figure 5.13) (McGroarty, 2010). However, if the vehicle pulls out of its path for any reason and enters the sidewalk, such a buffer cannot prevent it from hitting pedestrians or buildings due to its high speed and huge mass, but it does, however, increase the perceived safety and the sense of security among users.

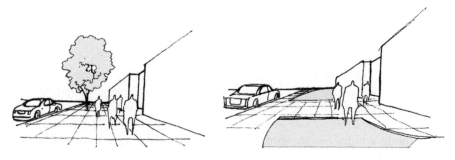

FIGURE 5.12 Appropriate design (left) and inappropriate design (right) based on the direction of the sidewalk

Source: Lotfalinejad, 2019

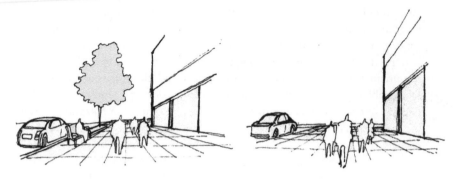

FIGURE 5.13 Appropriate (left) and inappropriate (right) design pattern based on creating a buffer between the sidewalk and the street

Source: Lotfalinejad, 2019

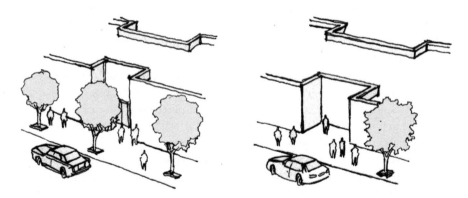

FIGURE 5.14 Appropriate (left) and inappropriate (right) design pattern based on the green space on the sidewalk margin

Source: Lotfalinejad, 2019

d) **Creating a pleasant green space on the sidewalk margins.** The proper width of the sidewalk alone cannot encourage citizens to walk. The green space on the margins of the sidewalk can have a significant effect on its desirability. The use of appropriate species of plants and trees (Figure 5.14) can moderate the atmosphere and leave positive climatic effects, such as increased shading in warm seasons and reduced wind speeds in the autumn and, to some extent, will serve as a temporary shelter during the rainy and cold seasons (CLADCP, 2008).

e) **Create a coherent rhythm.** Creating a visual rhythm in the sidewalk space has a tangible impact on the visual appeal of the sidewalk and the presence of people there. In its pedestrian-oriented "Urban Design Instructions", Niehoff Studio points out that elements with a vertical rhythm like light posts or trees can be used to provide visual appeal around sidewalks (Figure 5.15). This pattern of vertical repetition improves the relaxation of the human mind as, on the one hand, the human eye considers the repeating body a benchmark and can use it to recognize the beginning and end of the path. On the other hand, coherence in a repeating rhythm can be effective in reducing brain activity and preventing mental boredom on the part of the viewer in relation to a constantly changing environment. Policies for this goal are described in Table 5.11.

FIGURE 5.15   Appropriate (left) and inappropriate (right) design pattern based on the consistent rhythm of the sidewalk

Source: Lotfalinejad, 2019

TABLE 5.11 Summary of goals, strategies and policies focusing on the creation of walkable communities

Objectives	Strategies	Policies
Creating walkable communities	Proper mix of uses and density	– Convenient mix of uses to reduce the length of trips and turn walking into an efficient and convenient option – Quantitative and qualitative increase of walkability by placing uses adjacent to each other and creating multi-purpose destinations
	Zoning based on form	– Shortening the height of buildings and reducing the level of land occupancy as we move from neighbourhood centres to edges – Ensuring the proximity of different types of buildings in a neighbourhood unit – Ensuring the relative parallelism of the streets – with the exception of rare cases – and constructing similar types of buildings on both sides of the street – Changing zoning in the middle of the blocks and along the back edge of the land plots

(*Continued*)

TABLE 5.11 (Continued)

Objectives	Strategies	Policies
	Improving the existing streets and pavements to create pedestrian-oriented neighbourhoods Improving the convenience and safety of the sidewalks	– Targeted use of government and municipal funds in the transportation sector – Using a mix of bike and walking as the main transportation means – Investing in projects that enhance the safety and performance of biking and walking on daily trips – Creating appropriate sidewalks in a way that is both cost-effective and responds to pedestrians' needs (and are as straight as possible) – Establishing a buffer between the sidewalk and street using the natural landscape or street features – Creating favourable green space on the sidewalk using appropriate species of plants and trees – Creating a coherent rhythm, like a vertical rhythm through the use of light posts or using a coherent pattern of trees

Establishing basic services near residential areas, workplaces and public transport routes

Cities with medium and high densities and mixed land use usually have closer origins and destinations and create more incentives for citizens to walk. Research shows that density is of high importance in walking and the type of transportation means selected. Higher densities and mixed uses mean more crowds and more people moving within a walkable distance of public transport stations. It also means that streets with more people are more active, vibrant and secure. These contribute to the fact that developments with mixed uses and higher densities tend to increase walking and decrease the use of personal cars (Obeng & Ugboro, 2003). However, local zoning ordinances for new and infill developments in some cases prohibit and restrict the deployment of public services and people within walking a distance from home and workplaces. In addition, many neighbourhoods do not have (or lag behind) street standards that provide the required connectivity between neighbourhoods with mixed uses, street networks and existing public transport paths.

There are various mechanisms to deal with these obstacles. First, municipalities can recognize areas with mixed uses or the potential for mixed uses and then pay more attention to these areas in order to promote the dense nature of new and infill developments there (Berechman, Ozmen, & Ozbay, 2006). In addition, community groups and local governments can ensure the connection between streets and sidewalks to enhance the role of sidewalks. Ensuring that street building standards

TABLE 5.12 Summary of goals, strategies and policies focused on the deployment of basic services near residential areas, workplaces and transit routes

Objectives	Strategies	Policies
Establishment of basic services near housing, workplaces and transit routes	Utilizing mixed-use and high-density construction	– Accommodating the population within a walking distance of public transport stations – Identifying mixed-use areas – Promoting the dense and mixed-use nature of new and infill developments – Maintaining and enhancing the connections among streets and sidewalks – Emphasizing the importance of pavement in street standards
	Transit-oriented development	– Reconsidering transit bus routes as necessary and coordinating bus schedules in order to maximize the number of users – Establishing links among bus and subway stations to encourage more walking activities between public transit stations and destinations

confirm the importance of sidewalks can increase the walkability of any neighbourhood. Municipalities can reconsider bus routes and coordinate bus schedules to maximize the number of users. They can also connect bus stations to subway stations. As a result of such actions, more walking activities are expected among transit stations and destinations. Transit-oriented development seems to provide a key opportunity for the growth of new developments in the vicinity of existing neighbourhoods and allow more focus on walking options (Handy, 2005). Following Table 5.12 is the summary of policies and strategies for establishment of basic services near residential, commercial areas and transit routes.

Designing pedestrian-oriented commercial areas

Malls, offices, public facilities and other non-residential land uses are often destinations for urban trips and can be assets for cities. Varied street spaces with shops, restaurants, public arts and other uses encourage people to move slowly. A lively street has a good perspective and is safe and attractive (Figure 5.16). But a quiet street, which discourages walking activities, will soon become deserted and insecure. Some aspects of buildings that isolate people and prevent them from using sidewalks include faceless buildings (without doors and windows), or buildings without retail activities on the ground floor or streets that have a very low sidewalks width. Increasing the passage of people on the sidewalks of these areas requires

FIGURE 5.16   Assigning the first floor of buildings to stores in Lancaster

Source: Soeharjono, 2010

proper construction design to create a sense of place and security (Newman & Kenworthy, 2006).

Many design guidelines and instructions can be used by municipalities to make commercial areas pedestrian oriented. New construction guidelines include such items as the optimal separation of building components and blocks, the assignment of the ground floor of buildings to active uses, the rational visual permeability of building facades, incentives for the construction of sidewalks and the removal of parking between buildings and sidewalks.

In the streets that are part of the main structure of the vehicle network and have been designed with the aim of fast vehicle movement and access, the social role is reduced compared to other roles and vehicles that become more important than pedestrians. In such streets, if vehicles are removed, the connection between various spaces and the traffic network of the city faces a major problem. Therefore, in streets in which the movement or access role is more important than the social role and is part of the main roads used for vehicle movement and access to urban contexts, it is not possible to transfer their traffic to another street, and they cannot be turned into footpaths (Galingan, 2009).

In addition, in assessing the feasibility of transferring traffic to the surrounding streets, we should note that access to vehicles should be maintained for all residential units, health centres, parking lots, firefighters and any kind of use and activity

TABLE 5.13 Summary of goals, strategies and policies with a focus on designing pedestrian-oriented commercial areas

Objectives	Strategies	Policies
Designing pedestrian-oriented commercial areas	Creating a sense of place and security	– Encouraging people to walk through the creation of diverse street spaces with shops, restaurants, public arts and other uses – Preventing the construction of faceless buildings (without doors and windows) – Allotting the ground floors of buildings to stores and retail activities – Reducing the size of blocks – Removing parking between buildings and sidewalks – Maintaining access to vehicles for all residential units, health centres, parking lots, firefighting and any other types of users and activities that need access to cars in the feasibility studies of traffic transfer to surrounding streets

that requires access to cars. Finally, Table 5.13 shows strategy and relative policies for designing pedestrian-oriented commercial areas.

Providing safety for pedestrians and non-motorized vehicles

One of the factors influencing the walkability of neighbourhoods and urban areas is a sense of safety for pedestrians and cyclists in their daily trips. Pedestrians and cyclists should feel safe about sharing the streets with cars. Nevertheless, the design of today's streets is more focused on cars and places them at the top of the hierarchy of urban transport. But traditional street design has far more advantages compared to the modern one in terms of creating a sense of security and convenience. Smaller lots, narrower widths and street parking have created a balance in the traditional streets among different vehicle types. Moreover, they distribute street traffic and, given the low speed of cars, provide for pedestrian safety (Baumgartner et al., 2007).

The streets in neighbouring units and the city as a whole should be designed in such a way as to facilitate pedestrian crossings. For example, pedestrians will use pedestrian crossings if they do not have to go long distances for this purpose. This space is more important on the main streets. When there are fewer pedestrian crossings, unpredictable movements will occur on the street by pedestrians (Harkey & Zegeer, 2004). Local governments can avoid additional costs by developing and improving street standards in line with design features suitable for walking. Developing laws and incentives that encourage traditional street design before construction can be effective in designing a type of street that can support walking and other forms of non-motorized transportation. Cities can use the laws and regulations of physical zoning to ensure that new developments include street parking, sidewalks, narrow streets, small blocks, grid streets and bike lanes. Zoning can also be used to reduce the need of street width expansion and also designing of human-scale buildings (Martin, 2006).

One of the main challenges that should be addressed in the initial steps to achieve the goal of promoting the role of pedestrians in the city is to solve the puzzle of the intersection of footpaths with vehicle traffic. It should be noted that in this case, vehicles have physical superiority; therefore, such intersections should be avoided as much as possible. The most basic solution is to cross one over the other. In most cases, the sidewalk surface is elevated to a higher level (positive balance), using the pedestrian bridge to pass over the road to the other side of the road or to another street. In other cases, this access is provided through underground crossings (negative balance), creating a pedestrian underpass. Each of these solutions has its own advantages and disadvantages. However, traffic-calming measures are necessary in places where the separation of the footpath and the street is not possible or the disadvantages of the two methods are negligible. In this way, as the vehicle speed drops, a peaceful relationship can be created between pedestrians and motor vehicles.

Increasing pedestrians' field of view

Increasing the pedestrians' visibility (Figure 5.17) can be achieved by raising the level of the vehicle path at the intersection and creating a curvature at the edge of

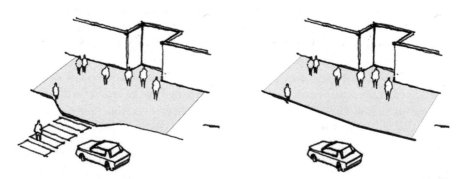

FIGURE 5.17   Appropriate (left) and inappropriate (right) design in terms of ensuring the safety of pedestrians and non-motorized vehicles on the sidewalk

Source: Lotfalinejad, 2019

the sidewalk (road setback in favour of the sidewalk). It is also necessary to remove marginal parking at the road edge at the intersection (CLADCP, 2008).

Reducing the length of pedestrian crossings on the streets

Create the shortest possible crossing distance at pedestrian crossings on wide streets (Figure 5.18). Devices that decrease the crossing distance may include a mid-street crossing island, an area of refuge between a right-turn lane and a through lane, a curb extension/bump-out and a minimal curb radius (CLADCP, 2008).

Although the pedestrian system has many benefits to the health of the city and its residents, the long distance between the origin and destination of urban trips sheds doubts on the effectiveness of this system. Due to physiological constraints, people usually do not walk distances more than 2.5 kilometres. Therefore, pedestrians should have good access to public transport on footpaths. This access can be in the form of a policy like "providing access to public transport stations at a 700-metre radius" (CLADCP, 2008).

Easy access for the disabled and underprivileged people to sidewalks, streets, parks and other public services

The enforcement of the Americans with Disabilities Act is not just a legal issue but the best way to satisfy the needs of all citizens (ADA, 2011). Although engineers are always struggling to meet the needs of people with disabilities when renovating streets, unfortunately, there is no easy technique to achieve this goal. Provision of pedestrian access and the enforcement of the previously mentioned act should be obligatory in all construction and reconstruction projects working on existing and new centres, streets and sidewalks (see Table 5.14). For example, the pedestrian crossings should be well designed to guide people to these places so that they can easily locate entry and exit points and avoid existing obstacles. Implementing the policies recommended in this section can lead to a significant increase in access for both healthy and disabled people and can meet the needs of people with disabilities without spending extra money.

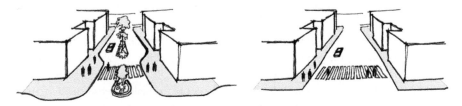

FIGURE 5.18   Appropriate (left) and inappropriate (right) design pattern in terms of reducing the length of pedestrian crossings on the streets

Source: Lotfalinejad, 2019

TABLE 5.14 Summary of goals, strategies and policies focusing on the safety of pedestrians and non-motorized vehicles

Objectives	Strategies	Policies
Ensuring the safety of pedestrians and non-motorized vehicles	Removing the sense of fear for pedestrians and cyclists in their daily trips	– Creating pedestrian crossings at appropriate distances – Taking into account factors such as on-street parking, sidewalks, narrow streets, small blocks, street grids and bike lanes zoning laws in new developments
	Calming road traffic	– Increasing the pedestrians' field of view by raising the road level at intersections and creating a curvature at the road edge (road setback in favour of the sidewalk) – Removing marginal parking on the road edge at the intersection – Reducing the length of pedestrian crossings on the street by creating islands, increasing the sidewalk surfaces in the form of curvatures and road refuge in the right-turn area – Providing access to public transport stations at a 700-metre radius
	Providing easy access to walkways, streets, parks and other public services for people with disabilities	– Enforcing the Americans with Disability Act in all new construction and reconstruction projects for new and existing centres, streets and sidewalks

Setting appropriate design standards for improving the quality of sidewalks

To meet the needs of pedestrians, sidewalks should have the appropriate width, buffer, continuity, connectivity and edges (Gerilla, Hokao, & Takeyama, 1995). Building design standards, public investment, regular assessment and proper maintenance can help make durable, convenient and safe sidewalks for citizens.

Connectivity is the primary goal of any transportation network and includes connection of places where people want to travel to. "Connectivity" refers to the straightness of travel routes and the number of routes available between each point (Heath et al., 2006). According to New Urbanists, the layout of streets in a neighbourhood to a large extent indicates the efficiency of the related transport system. New Urbanists emphasize the connection among streets as an important factor in encouraging walking, reducing driving and conserving energy (Kim, 2007). Accordingly, grid street pattern networks are preferred over street networks that

TABLE 5.15 Summary of goals, strategies and policies with a focus on designing appropriate sidewalk design standards

Objectives	Strategies	Policies
Setting appropriate design standards for improving the quality of sidewalks	Ensuring the proper width, buffer, continuity, connectivity and edges for sidewalks	– Connecting the places where people want to travel to in order to encourage walking, reduce driving and conserve energy – Creating grid street networks instead of patterns that include a large number of deadlocks and large blocks – Setting appropriate sidewalks in the central business districts to encourage more comfortable walking and increase the number of sidewalk users – Using sufficient trees to create a good landscape and protect the sidewalk – Placing sidewalks near buildings – Providing adequate light – Spatial separation of the sidewalk

include a large number of deadlocks and large blocks (cul-de-sac pattern) that increase the distance between destinations.

Design standards should specify the minimum sidewalk width, the required buffer to protect the users against cars and the minimum edges to reveal the sidewalk boundaries (Stockholm City Council, 2010). Using enough trees to create a good landscape and protect the sidewalk against traffic, placing sidewalks near buildings and providing adequate light and appropriate spatial separation in sidewalks will improve their quality and performance. In order to find appropriate design of standards for improving the quality of sidewalks, Table 5.15 summarized relative strategy and policies.

Calming traffic in residential neighbourhoods

Some of today's modern streets have been designed based on achieving the maximum speed of vehicles. Large lots, wide turning radii and wide streets encourage drivers to drive at high speeds, prevent pedestrians from moving, increase unsafe walking paths and, consequently, reduce the use of sidewalks. Traffic-calming techniques can balance the number of pedestrians and cars and help reduce traffic speed in neighbourhoods and main streets and encourage walking. Traffic-calming techniques can be used both for renovating existing streets and for designing new ones (Ewing, 1999).

Traffic-calming techniques generally include a variety of changes in street design such as installing speed bumps, narrowing the streets or arches inside the road to reduce the field of view, increasing the width of the sidewalks, designing non-straight roads, etc (see Table 5.16). These structural changes are often more effective

TABLE 5.16 Summary of goals, strategies and policies focusing on calming traffic in residential neighbourhoods

Objectives	Strategies	Policies
Implementing traffic-calming techniques in residential neighbourhoods	Using traffic-calming techniques to balance pedestrian and car numbers	– Designing intersections based on pedestrian priority through traffic signs and surface colour changes – Creating speedboats in crowded areas – Narrowing the streets or curves inside the road to reduce the field of view – Increasing the width of the sidewalks to encourage walking – Designing non-straight roads

ways of calming traffic compared to compulsory methods and helps with different types of transportation such as bikes, cars and buses alongside pedestrians (Williams, Seggerman, & Nikitopoulos, 2004).

Protecting and beautifying existing and new sidewalks

The construction of pedestrian-oriented cities not only means the construction of sidewalks and other pedestrian equipment for citizens (such as pedestrian cross-walks, bike lanes, sidewalks, etc.), it also means protecting and maintaining (Figure 5.19) these infrastructures (Teir, 1993). Sidewalks and streets that are not well protected and maintained are unacceptable for walking and are a serious threat to the safety of cyclists and people using non-motorized vehicles for their trips. Attractive and protected walkways encourage more people to walk on their trips. Cities that create scenery and landscape around roads, urban centres, open spaces and sidewalks use these attractive environments to further encourage walking. For example, Birmingham in Michigan has seen a drop in traffic speed from 10 to 15 km/hr on the streets with trees. The use of public art, places to sit and the presence of waste bins in busy urban centres like downtowns, squares and main parks, as well as along the main transport corridors, can enhance the walking experience (Boyld, 2006).

Sidewalks should therefore be maintained so that they can provide a hospitable pedestrian environment and their useful life increases (Bannister, Fyfe, & Kearns, 2006). Snow removal on sidewalks should be done quickly and, if necessary, their slabs repaired or replaced. Periodic repairs and continuous maintenance of sidewalks are indispensable. Healthy cities should recognize and investigate maintenance problems well. Shrubs, trees and other plants should be decorated regularly (see Table 5.17 for relative policies). Local departments can develop and implement building codes and standards needed to maintain footpaths and thus help property owners increase their investment in building safer and more secure footpaths (Boyld, 2006).

FIGURE 5.19 Protecting and maintaining the sidewalk can add to the attractiveness of walking

Source: SPUR, 2011

TABLE 5.17 Summary of goals, strategies and policies with a focus on protecting and beautifying existing and newly built sidewalks

Objectives	Strategies	Policies
Protecting and beautifying existing and newly built sidewalks	Building attractive and protected sidewalks in order to encourage more people to walk to reach their destinations	– Creating a beautiful landscape and scenery (such as trees and plants) around roads, urban centres, open spaces and sidewalks – Use of public art like street paintings, music performance, etc. along the pavement to increase the attractiveness of walking – Putting seats and waste bins in urban centres such as in downtowns, squares and main parks of the city – Periodic repairs of pavement and permanent maintenance of sidewalks – Use of stop signs for calming traffic speed – Rapid snow removal of sidewalks during snowfall – Adorning and decorating shrubs, trees and other plants regularly

The presence of attractive edges on the street

Attractive edges are of particular importance in creating footpaths, because they create life and experience on the streets. In order to have a lively and vibrant city, a large part of the edges of the street should be open, transparent and pleasant to create a sense of safety and beauty for movement and circulation in the city (Gehl, 2002). Therefore, streets that become footpaths should have an attractive edge, or at least a pleasant edge. Table 5.18 shows the classification of street edges based on the degree of attractiveness. Additionally, Table 5.19 demonstrates strategy and policies related to creating attractive street edges.

TABLE 5.18 Classification of street edges in terms of attractiveness

Attractive (A)	Small units with many doors (15 to 20 units per 100 metres); various uses; lack of closed or inactive units; attractiveness of the facades; quality materials and details
Pleasant (B)	Relatively small units (10 to 14 units per 100 metres); relatively different uses; few units closed or inactive; relative attractiveness of the facades; relatively good detail
Medium (C)	A combination of large and small units (6 to 10 units per 100 metres); relatively few uses; only a few store units closed or inactive; unattractive facade design; relatively poor details
Boring (D)	Large units with relatively small doors (2 to 5 units per 100 metres); low diversity of uses; lots of units closed; many unattractive facades; little or no detail
Unattractive (E)	Large units with few or no doors; lack of visible variation in uses; inactive or closed facades; boring facades; lack of detail; nothing interesting to watch

Source: Gehl & Soholot, 2002

TABLE 5.19 Summary of goals, strategies and policies focusing on creating attractive street edges

Objectives	*Strategies*	*Policies*
Creating attractive street edges	Creating open, transparent and pleasant street edges	– Creating small units with many doors (15 to 20 units per 100 metres) at the edge of streets – Using street edges for different uses – Preventing the creation of closed and passive units at the street edges – Using high-quality and local materials with desirable details to increase the attractiveness of the facades

The presence of voluntary and social activities on the streets

According to Gehl and Soholot (2002), there are three groups of outdoor activities in urban spaces: necessary, optional and social activities. In low-quality areas, only necessary activities, which people are forced to do (such as commuting), can be seen. In high-quality urban areas, in addition to necessary activities, people engage in many recreational and social activities that are of interest to the public. However, these activities occur when conditions are appropriate, that is, the city provides quality, tempting spaces. Therefore, a good city can be compared with a good party (Figure 5.20): people stay there longer than what is really needed, because they enjoy being there (Gehl & Soholot, 2002, p. 9). So, the more a street has social activities, the more suitable it will be for becoming a footpath. Additionally, Table 5.20

FIGURE 5.20 Nightlife in Istiklal Street, İstanbul

Source: Tunda, 2007

TABLE 5.20 Summary of goals, strategies and policies focusing on the availability of optional and social activities at the street level

Objectives	Strategies	Policies
Presence of optional and social activities on the street	Creating quality and appealing spaces for attracting people to stay in urban spaces	– Increasing the number of social activities in the streets – Increasing the number of round-the-clock uses in crowded areas

shows two policies for creating quality and appealing spaces for attracting people to stay in urban spaces.

Moreover, streets with evening activities have a high potential for turning into footpaths, because "the number of night activities and their location is an important factor for city vitality and perception of safety. If the number of activities at night is low, visitors will feel that the city is empty and they avoid going out" (Gehl & Soholot, 2002, p. 41).

Increasing the readability of urban routes

Visual and physical permeability

Cities require a lot of connections to facilitate walking. Even when residential and commercial areas are adjacent to each other, without proper connections, citizens will not be able to replace short car trips with walking. Unfortunately, in many cases, the network design and the distribution of land use in a zone is done without regard for the principles of connectivity and permeability, whereas traditional street networks that had both smaller and larger segments often provided more routes to reach the destination (Morris, 2013). In this regard, the renovation of existing street networks in such a way as to create strong connections with traditional street networks, valuable corridors, parking lots, greenways, footpaths and other open spaces would be very beneficial.

As a result, walkways should have high accessibility and permeability for the public. The permeability of each system depends on the number of potential routes that are available for going from one point to the other. In addition, visual permeability is also important, meaning that the streets must be transparent and visible (Montgomery, 1998); otherwise, only those already familiar with the area can benefit from it. So, the more paths a street is connected to, the more physical and visual permeability it will have.

Proper utilization of existing urban signs

Elements with special characters along, at the beginning or at the end of the paths not only enhance street readability (Figure 5.21) but also increase its appeal and invite people to the place. They lead to increased activity and the use of the street, as people can experience the environment and have many opportunities like shopping, entertainment and more. Therefore, the streets with more signs and descriptive elements attract more pedestrians (Lynch, 1981). In this regard, paying attention to the optimal number of signs and the visual and physical connection among them, along with their correct placement and distribution in the design area, can help to optimize the performance of these elements. Accordingly, Table 5.21 is the checklist of policies and strategies for improving the readability of urban routes.

FIGURE 5.21 An urban sign, which is in fact part of the street installation art, has become a tourist attraction in addition to increasing readability and creating a distinctive character for the street

Source: Elcorredor, 2009

TABLE 5.21 Summary of goals, strategies and policies with a focus on increasing the readability of urban routes

Objectives	Strategies	Policies
Improving the readability of urban routes	Visual and physical permeability	– Increasing the visibility and transparency of urban roads and paths – Connecting valuable corridors, parking lots, greenways, footpaths, open spaces and other key uses together – Fine-grain zoning of urban blocks to the optimum level
	Proper utilization of urban signs	– Placing urban signs at the beginning or the end of roads or alongside them – Considering the number of optimal signs – Ensuring the existence of visual and physical connections between urban signs – Ensuring the correct location and layout of urban signs – Distributing symbols on the design area in an appropriate way

Understanding the economic opportunities that encourage activity on footpaths

Design standards, traffic control techniques and other policies discussed in this section help create an environment that is appropriate for better walking activity. In addition to the direct mechanisms for creating pedestrian-oriented cities, municipalities can identify economic and business opportunities that encourage walking. Redevelopment plans for main streets, first-floor stores, sidewalks and shopping centres or pedestrian markets are ways to develop footpath activities for the purpose of economic development and capitalization. Recognizing important local assets such as natural features, historical areas or unique architectural designs can enhance the walkability of the city. Economic development strategies that utilize such assets can be implemented to attract pedestrians for retail and restaurant activities and thus increase walking. For example, many cities have plans for main streets that are used to revitalize urban cores or downtown corridors. However, in some small towns in New York, the revitalization of main historic streets has been done with the aim of attracting tourists to urban centres. This will encourage people to walk in old shops, crafts and other local shops in this part of the city (Zukin, 1987).

Other cities like Boulder, Colorado; Burlington, Vermont; and Charlottesville, Virginia, have been attracting tourists by building department stores in the specific walk streets which presence of vehicles are prohibited (Figure 5.22) (Campbell,

FIGURE 5.22 Creating footpaths and revitalizing shops in a business area to attract pedestrians

Source: Authors

TABLE 5.22 Summary of goals, strategies and policies with a focus on recognizing economic opportunities that encourage activity on footpaths

Objectives	Strategies	Policies
Recognition of economic opportunities encouraging activities on footpaths	Creating economic and business opportunities to encourage mobility and walking	– Establishing retail stores and restaurants on the ground floor to attract pedestrians – Revitalizing historic streets with the aim of attracting tourists to urban centres – The establishment of old stores, handicrafts and other shops for the supply of cultural products in the city's historic streets in order to attract tourists – Creating shopping areas inside the sidewalks in order to attract tourists and increase the vitality of the space – Creating a walkable commercial space with a two-way road in the main street of the city centre – Closing down the main streets of downtown to vehicles to hold art fairs and other special events on some days

2013). As the number of customers who can access the shops on foot increases, more shops are built to attract them. Thus, cities can use economic revitalization as an appealing force to expand walking activities in different ways.

In some cities across the world, on certain days of the year, downtown streets are closed to cars for the purpose of holding art fairs and other special events, leading to vibrant urban spaces throughout the day and night (Bendick Jr & Rasmussen, 1996).

Finally, Table 5.22 describes six policies for better recognition of economic opportunities which encouraging activities on footpaths.

Creating an equilibrium between development and environmental protection

Although the footprints of the discussion on the city's relationship with nature can be seen in the critiques of modernist urbanism from the 1930s (Mumford's *Brown*

Decades), with the release of Rachel Carson's *Silent Spring* in 1962, American society's attention was drawn to the environmental issue of cities, and this book could have a significant impact on the increasing importance of environmental issues. On the other hand, with the increasing population of cities, urban planners began to pay more attention to the urban environment. In the meantime, smart growth is seeking to achieve a dynamic and growing environment, which itself can leave a valuable legacy to later generations. Smart growth challenges the methods of construction, work and life. This kind of development forces us not to consider urban areas merely as places to live but to consider them as a means to promote health and achieve prosperity and a desirable life. As discussed in previous sections, smart growth is considered one of the most innovative solutions to urban problems and it tries to "balance development and environmental protection" (U.S. Environmental Protection Agency, 2002).

In order to achieve this goal, two strategies have been suggested: "protecting sensitive areas of the environment" and "achieving a dynamic and growing environment". There are various policies for the realization of each one. The main policies for achieving the first strategy include reducing consumption, protecting open spaces, protecting particular species of animals and plants, protecting water resources, maintaining good air quality and maintaining land located on the edge of urban areas and sensitive ecology.

Some other policies can be effective in realizing the strategy of achieving a dynamic and growing environment, for example, the policy of protecting open spaces and sensitive environments in and around cities and towns will protect the natural lands around cities against urban trespass and urban development. The policy of using surface runoff can have many benefits in creating drought-tolerant landscape plans and the creation of a recycling process. It can be used to irrigate green spaces and return water to the groundwater cycle and, to a lesser extent, can be used in carwashes and the like. Whenever possible, natural landscapes and plants should be protected and the best natural potentials should be deployed in green parks and green belts.

On the other hand, providing adequate public facilities in development areas, under the supervision of local management, reducing the use of intact land and natural resources and reducing the pollution in rivers and lakes help protect the natural environment and make the community more attractive, more economic, more diverse and stronger. Table 5.23 demonstrates policies for two strategies in order to balancing development and environmental protection.

TABLE 5.23 Summary of goals, strategies and policies focused on balancing development and environmental protection

Objectives	Strategies	Policies
Balancing development and environmental protection	Protecting major environmental capacities	– Reducing consumption of resources – Protecting natural and open spaces

Objectives	Strategies	Policies
	Achieving dynamic and growing urban spaces within cities	– Protecting particular species of animals and plants – Protecting water resources – Maintaining land located on the edge of urban areas and areas with sensitive ecology – Controlling wind and reducing air pollution – Reducing noise – Controlling erosion and temperature fluctuations – Protecting open spaces and sensitive environments in and around cities – Utilizing surface runoff – Using drought-tolerant landscape plans – Using drainage and treatment of wastewater, wastes and chemical contaminants and creating a water recycling process – Deploying the best natural potential in parks and green belts – Providing adequate public facilities and equipment, under the supervision of local management, in development locations

Proper utilization of various uses

Jacobs (1961) states that the complex mix of different uses in cities in no way represents a form of chaos, but, on the contrary, is itself a manifestation of a complex and highly developed form of order (Seifuddini & Shorjah, 2014). Based on this approach, a mix of various commercial, residential, recreational, educational and other uses is considered the most important way of creating a dynamic and social vitality and a sense of place (HosseinZadeh Delir & Safari, 2012). In fact, when diverse and varied uses come together, more people will be present in the streets, resulting in increased vitality of urban spaces. On the other hand, due to shorter paths for access to different uses, more people will be attracted to the area and there will be the possibility to compress urban contexts. Also, with the centralization and proximity of uses in an area, savings will result from aggregation, helping to boost the economies of various areas. Accordingly, this type of development encompasses many benefits and can lead to greater vitality, sustainability, socialization, appropriate accessibility, increased security, increased social interactions and more effective use of the existing infrastructure. Given these benefits, mixed use has become one of the most important goals of smart growth.

One of the most important strategies is the creation of mixed uses with different functional scales in population centres. Different policies can be used to achieve this strategy. Combining business, civic, cultural and recreational uses in the central focus of each community is a policy that itself encourages pedestrian access to a range of activities located in the related centre of activity. Another policy that can be effective in reducing the number of city trips is the establishment of shopping centres for city centres and towns in such a way that they can be accessed only by a single parking lot.

Another strategy is stressing mixed uses in urban neighbourhoods. To this end, public spaces and facilities for daily use in urban neighbourhood structures can be effectively used. This policy provides various options for determining a building's use for developers and residents of neighbourhoods, and only the physical and aesthetic features of public spaces are driven by city administrators (Reese, 2011).

Another policy that can be used in conjunction with the mixed-use strategy is that of placing shopping malls, schools and businesses in the vicinity of residential neighbourhoods. This policy can lead to a reduction in the length of trips among a group of important urban activities that are considered major destinations for urban trips. As a result of shorter distances, less air pollutants and greenhouse gases are released in the neighbourhoods and metropolitan areas (Filion & McSpurren, 2007; Miller & Hoel, 2002).

Other methods of achieving mixed uses include modifying and reviewing the city codes and regulations on urban uses and placing such uses in various zoning plans or urban planning plans; to this end, we can adopt policies for providing financial incentives, encouraging redevelopment in single-use areas and using improved zoning techniques (such as flexible zoning, overlay zoning and impact zoning). Table 5.24 represents relative policies to two strategies under adopting mixed uses to enhance visibility opportunities.

TABLE 5.24 Summary of goals, strategies and policies focusing on the proper utilization of mixed uses

Objectives	Strategies	Policies
Adopting mixed uses to enhance visibility opportunities	Creating diverse uses with different functional scales in population centres	– Mixing business, civic, cultural and recreational uses together in the central focus of each community – Establishing shopping centres for city centres and towns in such a way that they can be accessed only by a single parking lot

Objectives	Strategies	Policies
	Paying attention to mixing uses in urban neighbourhoods	– Establishing public spaces and facilities for daily use in the urban neighbourhood structure – Locating shopping centres, schools and businesses near residential neighbourhoods
	Modifying and revising the rules for the deployment of urban uses	– Providing financial incentives to create mixed uses – Encouraging re-development in single-use regions – Using improved zoning techniques (such as flexible zoning, overlay zoning and impact zoning).

Preventing urban sprawl: moving towards planning and designing a compact city

As mentioned in the previous chapters, urban sprawl can lead to various costs and damages, which can be divided into two general categories (Carruthers & Ulfarsson, 2002; Litman, 2015b): The first is per capita land consumption costs. Undoubtedly, the costs of public service and facilities also rise with an increase in land consumption. As a result, more agricultural land is lost, food production decreases and natural resources are damaged, and the cost of offsetting or reducing these damaging effects is also added to the major costs of urban sprawl.

The second category includes the costs of dispersion of activities, which in turn, reduces access across the city and makes the creation of a public transport infrastructure more difficult and less cost-effective. This activity dispersion also reduces the possibility of creating pedestrian- or bike-oriented transport systems. Also, by relying more on car-based transportation, the city faces challenges such as parking, traffic jams and accidents, which impose costs and heavy pressure on the public sector and taxpayers (Litman, 2015a).

On the other hand, according to some experts, allocating land to intensive development as one of the methods for sustainable land use planning "is in line with creating urban dynamism and equity for the general public by increasing citizens' access to urban uses" (Berke, Godschalk, Kaiser, & Rodrigerz, 2006). According to activists of the Trust of Public Land, the most important point for achieving smart growth is that more people believe that life in smaller urban areas with denser development can result in increased liveliness. The organization also believes that growth will be truly smart if the human community decides which parts of the land will be saved to preserve the natural resources and open spaces in the settlements. Such a decision helps smart growth and compact development occurs (Trust for Public Land, 1999).

Three strategies can be helpful in addressing the widespread social, economic and environmental issues of urban sprawl: reducing the consumption of land and natural resources, directing development towards central and inner-city lands and controlling the unrestrained and disordered growth. If less urban land and natural resources is used, we can expect a reduction in the need for infrastructure, roads and parking, as well as in maintenance costs, and ultimately an increase in opportunities to save and reimburse the costs of infrastructure integration and transportation. For this purpose, two policies can be implemented: specifying the distinct boundary between suitable land for development and lands that need protection and creating green belts of agriculture or natural paths. Several policies can be used in line with the strategy of guiding development towards central and inner-city urban areas. For example, the optimal and maximum use of existing infrastructure and facilities within cities for development and increasing the awareness of individuals regarding the economic, environmental and social benefits of the development in central regions and the allocation of legal incentives are key policies for directing economic capital towards internal areas of the city and to reuse and optimize the old urban contexts. Ultimately, to control disordered and excessive development, policies for empowering, reinvigorating, and revitalization of new cities and towns and low-density suburbs and reusing brown fields can be implemented. These policies, to a large extent, can prevent the horizontal expansion of most cities and the destruction of farmland and other pastures.

On the other hand, as noted, compact urban planning and design is considered to be one of the main goals of smart growth in order to improve residents' quality of life and ultimately achieve sustainable development. In order to achieve this goal, two strategies can be used: supporting optimal dense development and dense allocation of uses. The first strategy is used to reach a denser urban area with more mixed uses and gives special attention to the structural aspect of building density. In this regard, we can use the policy of "designing infrastructure, transportation system and, most importantly, land use system with optimized density" to create places that have more appeal and a greater sense of place.

Some of the city's problems relate to parking requirements and lack of awareness regarding the benefits of dense development. In order to solve these problems, neighbourhoods can create more compact plans. For example, instead of large buildings, the policy of building multi-story parking structures can be implemented. Another policy that can be used in line with this strategy is constructing housing with optimal density, which can be achieved through a mix of single-family and multi-family houses in existing neighbourhoods.

Another strategy used to achieve the goal of compact urban design and planning is the correct location of urban uses, which supports public transport, as well as pedestrian- and bicycle- oriented transportation. An important policy in the pursuit of this strategy is that of locating institutions and organizations with the urban and regional level of service in the central cores of cities. The policy of deploying essential services at close distances can also reduce dependency on cars and the demand for parking spaces.

TABLE 5.25 Summary of goals, strategies and policies with a focus on preventing urban sprawl and moving towards dense urban design and planning

Objectives	Strategies	Policies
Preventing urban sprawl	Less consumption of lands with natural value	– Specifying the distinct boundary between suitable lands for development and lands that need protection – Creating green belts of agriculture and natural paths
	Guiding the development towards the central and inner-city areas	– Identifying suitable sites for infill development with optimal density – Optimal and maximum use of infrastructure and facilities available within cities for development – Increasing people's awareness of the economic, environmental and social benefits of developments in central urban areas – Allocating legal and financial incentives
	Controlling the disordered and unrestrained growth of development	– Empowering, reinvigorating and revitalizing new cities and towns and low-density suburbs – Reusing brown fields
Creation of concentrated and compact urban communities	Supporting development with optimal density	– Considering optimum density in designing infrastructure, transportation systems and land use – Constructing multi-story parking structures – Constructing housing with optimal density and mixing single-family and multi-family houses
	Correct location of uses	– Locating institutions and organizations with the urban and regional level of service in the central cores of cities – Establishing essential services at close distances to reduce car dependency and demand for parking spaces

In the end, it can be said that smart growth causes the development of areas to be carried out with an appropriate level of use density, which will reduce the use of personal vehicles (Girardet, 2004). Description of objectives and its relative strategies and policies for this section can be found in Table 5.25.

Summary of generalizable and achievable goals, strategies and policies for smart growth (with an emphasis on pedestrian-oriented transportation)

In this section, based on the previous sections, a summary of operational goals, strategies and policies is presented in Table 5.26.

TABLE 5.26 Summary of generalizable and achievable goals, strategies and policies for smart growth (with an emphasis on pedestrian-oriented transportation)

Objectives	Strategies	Policies
– Reducing reliance on personal transportation – Providing a variety of transportation options – Reducing the consumption of fossil fuels and, consequently, their pollutants, such as carbon derivatives and greenhouse gases – Promoting people's health	– Increasing compatibility between land use and the public transport system – Increasing the number of transportation vehicles – Increasing access to high-quality transportation services – The quantitative and qualitative promotion of public transport Creating and developing bike systems	– Increasing the number of active buses in the urban bus system – Increasing the number of subway lines – Suitable climatic design for bus and subway stations – Informing people of the schedules of buses and subway trains – Presenting online travel information for travellers so that they can better plan for the use of the public transport network – Improving the quality of public transport and equipping vehicles with heating, cooling and ventilation equipment – Using NGVs in the public transport system – Equipping public transport stations with solar energy systems – Combining the bike system with other types of public transport – Allocating bike lanes with a suitable width between pedestrian paths and streets – Maintaining and enhancing the safety and security level for users of this transport system using measures such as the use of a shield or buffer along bike paths. – Using smart light control systems at intersections – Using smart designs for the shape and form of bike paths and avoiding designing circuitous paths as much as possible (especially in the paths leading to intersections) – Using the shortest routes for the bike system to link the main functional nodes of the city or the main centres of travel production – Creating attractive and pleasant bike paths by prioritizing the use of natural landscapes such as the margins of urban parks, valleys, canals, etc. Using green plant strips and planting climate-compatible plants or trees on the bike lane edges

Objectives	Strategies	Policies
Optimal use of land for targeted development in existing urban areas	Compatibility of the transportation system with urban development	– Implementing intensive and infill development policies – Redevelopment of brown fields and barren lands in order to take advantage of the capacity of existing transportation networks – Prioritizing projects that properly provide for mixing land uses in their design
	Strengthening local transport systems to adequately respond to the daily needs of user groups	– Providing incentives and grants to local administrations to support cross-sectoral investments in areas involving mixed uses – Providing incentives for urban areas with safe passageways for schoolchildren Providing incentives for neighbourhoods that have attracted crowds and key uses in walkable places
	Adaptation of use location with existing and proposed transportation systems	– Locating new uses or changing existing uses in the vicinity of existing transportation networks – Choosing the type and routes of transportation networks based on the performance of existing uses in urban areas – Functional performance overlapping of various urban areas and full adaptation of the transport network to existing areas – Mixing urban uses
	Paying attention to the optimal density of urban areas	– Optimal loading of building density in order to maximize the use of land in urban areas, thereby reducing the need for new trans-shipping networks Paying more attention to the quality and diversity of urban transport networks, rather than to the quantity of areas assigned to them
Improving and modifying road service-level standards in public transport areas	Creating a balance between the road's service level and traffic flow	– Improving the level of service based on a local community's needs resulting from their activity and residence – Improving the bike level of service based on bike ownership in the area – Improving the personal car level of service based on the average daily traffic of personal cars

TABLE 5.26 (Continued)

Objectives	Strategies	Policies
	Upgrading the pathway to the level of service of a full street	– Providing a bike route – Providing a public transit lane – Providing a route for rail transport at the zero level – Providing sidewalks on the edges of the streets – Improving the economic vitality and prosperity in urban centres
Connecting local area networks (with two to four lines) to higher transportation levels	Creating fused grid street layouts out of small-scale streets	– Connecting the main streets of neighbourhoods to one another and to a higher level of transportation – Maintaining the internal tranquillity of neighbourhoods using impasses – Maintaining the role and size of auxiliary streets proportional to the neighbourhood level
Connecting different types of transportation networks to each other	Establishing connections among different public transportation systems	– Creating bike places at public transport stations – Creating sidewalks among different locations to enhance the efficiency of trips – Facilitating citizens' trips among bus and subway systems
	Making public transit stations closer to sources of urban trips	– Expanding public transit stations at short distances (up to a maximum of 300 metres) from one another – Increasing residential density around the stations Locating public transit stations within 400 metres of the residential are
Supporting public transportation services through land use	Creating an area or zone for activity centres around the transportation systems	– Developing residential buildings with higher densities around transportation stations – Locating services like childcare services, as well as facilities for daily trips, such as laundry and ironing facilities and convenience stores, along with transportation stations in order to increase the efficiency and attractiveness of public transport for citizens
Creating pleasant and attractive walkways in all new developments in the city	Mixing uses, small blocks and close destinations	– A balanced mix of transportation options – The requirement for the construction of sidewalks in new buildings by the government

Objectives	Strategies	Policies
Meeting parking needs	Optimize access to parking in local destinations	– Using street parking to provide the required stopping distance by municipalities – Reducing the number of parking spaces required for new construction in areas of public transport or mixed-use areas – Specifying priority areas for parking – Reducing the cost of parking for shared cars – Allowing non-cash payment of parking fees for employees – Allocating income resulting from parking taxes to workers – Encouraging builders to construct parking behind the building or inside the yard
Giving incentives to reduce traffic at peak hours	Giving information and incentives for commuting options	– Integrating different departments and offices at locations close to subway stations – Providing housing in partnership with employers, for employees who want to live close to their workplace – Providing the opportunity to work at home through teleworking – Providing flexible working hours through flexible times
Establishing coordination among different transport services	Creating land use compatible with public transport	– Placing bus stations or routes in high-density areas – Locating stations in areas with supporting land use – Locating bus stations in mixed-use areas with access to high-quality walking paths – Setting transport system goals based on the population and economic factors of the target community – Considering the nature of trips and travel patterns and making decisions accordingly
	Providing a wide range of high-quality housing options for all income groups	– Changing the structure of single-family housing and expanding housing options, including flats, residential complexes and multi-family housing – Combining single-family and multi-family housing in existing neighbourhoods

(*Continued*)

TABLE 5.26 (Continued)

Objectives	Strategies	Policies
		− Providing different housing options in various sizes and prices (in the form of duplexes, flats and single-family detached housing) in compact and dense contexts
		− Providing affordable housing for low-income households
		− Providing homes suitable for the elderly and disabled people
Creating walkable communities	Proper mix of uses and density	− Convenient mix of uses to reduce the length of trips and turn walking into an efficient and convenient option
		− Quantitative and qualitative increase of walkability by placing uses adjacent to each other and creating multipurpose destinations
	Zoning based on form	− Shortening the height of buildings and reducing the level of land occupancy as we move from neighbourhood centres to edges
		− Ensuring the proximity of different types of buildings in a neighbourhood unit
		− Ensuring the relative parallelism of the streets − with the exception of rare cases − and constructing similar types of buildings on both sides of the street
		− Changing zoning in the middle of the blocks and along the back edge of the land plots
	Improving the existing streets and pavements to create pedestrian-oriented neighbourhoods	− Targeted use of government and municipal funds in the transportation sector
		− Using a mix of bike and walking as the main transportation means
		− Investing in projects that enhance the safety and performance of biking and walking on daily trips
	Improving the convenience and safety of the sidewalks	− Creating appropriate sidewalks in a way that is both cost-effective and responds to pedestrians' needs (and are as straight as possible)
		− Establishing a buffer between the sidewalk and street using the natural landscape or street features

Objectives	Strategies	Policies
		– Creating favourable green space on the sidewalk using appropriate species of plants and trees
		– Creating a coherent rhythm, like a vertical rhythm, by using light posts or using a coherent pattern of tree.
Establishment of basic services near housing, workplaces and transit routes	Utilizing mixed-use and high-density construction	– Accommodating the population within a walking distance of public transport stations
		– Identifying mixed-use areas
		– Promoting the dense and mixed-use nature of new and infill developments
		– Maintaining and enhancing the connections among streets and sidewalks
		– Emphasizing the importance of pavements in street standards
	Transit-oriented development	– Reconsidering transit bus routes as necessary and coordinating bus schedules in order to maximize the number of users
		– Establishing links among bus and subway stations to encourage more walking activities between public transit stations and destinations
Designing pedestrian-oriented commercial areas	Creating a sense of place and security	– Encouraging people to walk through the creation of diverse street spaces with shops, restaurants, public arts and other uses
		– Preventing the construction of faceless buildings (without doors and windows)
		– Allotting the ground floors of buildings to stores and retail activities
		– Reducing the size of blocks
		– Removing parking between buildings and sidewalks
		– Maintaining access to vehicles for all residential units, health centres, parking lots, firefighting and any other types of users and activities that need access to cars in the feasibility studies of traffic transfer to surrounding streets
Ensuring the safety of pedestrians and non-motorized vehicles	Removing the sense of fear for pedestrians and cyclists in their daily trips	– Creating pedestrian crossings at appropriate distances
		– Taking into account factors such as on-street parking, sidewalks, narrow streets, small blocks, street grids and bike lane zoning laws in new developments

(Continued)

TABLE 5.26 (Continued)

Objectives	Strategies	Policies
	Calming road traffic	– Increasing the pedestrians' field of view by raising the road level at intersections and also creating a curvature at the road edge (road set back in favour of the sidewalk)
		– Removing marginal parking on the road edge at the intersection
		– Reducing the length of pedestrian crossing on the street by creating islands, increasing the sidewalk surfaces in the form of curvatures and road refuge in the right-turn area
		– Providing access to public transport stations at a 700-metre radius.
	Providing easy access to walkways, streets, parks and other public services for people with disabilities	– Enforcing the Americans with Disability Act in all new construction and reconstruction projects for new and existing centres, streets and sidewalks
Setting appropriate design standards for improving the quality of sidewalks	Ensuring proper width, buffer, continuity, connectivity and edges for sidewalks	– Connecting the places where people want to travel to in order to encourage walking, reduce driving and conserve energy.
		– Creating grid street networks instead of patterns that include a large number of deadlocks and large blocks
		– Setting appropriate sidewalks in the central business districts to encourage more comfortable walking and increase the number of sidewalk users
		– Using sufficient trees to create a good landscape and protect the sidewalk
		– Placing sidewalks near buildings
		– Providing adequate light
		– Spatial separation of the sidewalk
Implementing traffic-calming techniques in residential neighbourhoods	Using traffic-calming techniques to balance pedestrian and car numbers	– Designing intersections based on pedestrian priority through traffic signs and surface colour changes
		– Creating speed bumps in crowded areas
		– Narrowing the streets or curves inside the road to reduce the field of view
		– Increasing the width of the sidewalks to encourage walking
		– Designing non-straight roads

Objectives	Strategies	Policies
Protecting and beautifying existing and newly built sidewalks	Building attractive and protected sidewalks in order to encourage more people to walk to reach their destinations	– Creating a beautiful landscape and scenery (such as trees and plants) around roads, urban centres, open spaces and sidewalks – Use of public arts like street paintings, music performance, etc. on the pavement to increase the attractiveness of walking – Putting seats and waste bins in urban centres such as in downtowns, squares and main parks of the city – Periodic repair of pavement and permanent maintenance of sidewalks – Use of stop signs for calming the traffic speed – Rapid snow removal of sidewalks during snowfall – Adorning and decorating shrubs, trees and other plants regularly
Creating attractive street edges	Creating open, transparent and pleasant street edges	– Creating small units with many doors (15 to 20 units per 100 metres) at the edge of streets – Turning street edges to edges with different uses – Preventing the creation of closed and passive units at the street edges – Using high-quality and local materials with desirable details to increase the attractiveness of the facades
Presence of optional and social activities on the street	Creating quality and appealing spaces to encourage people to stay in urban spaces	– Increasing the number of social activities in the streets – Increasing the number of round-the-clock uses in crowded areas
Improving the readability of urban routes	Visual and physical permeability	– Increasing the visibility and transparency of urban roads and paths – Connecting valuable corridors, parking lots, greenways, footpaths, open spaces and other key uses together – Fine-grain zoning of urban blocks to the optimum level
	Proper utilization of urban signs	– Placing urban signs at the beginning or the end of roads or alongside them – Considering the number of optimal signs

(Continued)

TABLE 5.26 (Continued)

Objectives	Strategies	Policies
		− Ensuring the existence of visual and physical connections between urban signs
		− Ensuring the correct location and layout of urban signs
		− Distributing symbols on the design area in an appropriate way
Recognition of economic opportunities encouraging activities on footpaths	Creating economic and business opportunities to encourage mobility and walking	− Establishing retail stores and restaurants on the ground floor to attract pedestrians
		− Revitalizing historic streets with the aim of attracting tourists to urban centres
		− The establishment of old stores, handicrafts and other shops for the supply of cultural products in the city's historic streets in order to attract tourists
		− Creating shopping areas inside the sidewalks in order to attract tourists and increase the vitality of the space
		− Creating a walkable commercial space with a two-way road in the main street of the city centre
		− Closing down the main streets of downtown to vehicles to hold art fairs and other special events on some days
Balancing development and environmental protection	Protecting major environmental capacities	− Reducing consumption of resources
		− Protecting natural and open spaces
		− Protecting particular species of animals and plants
		− Protecting water resources
		− Maintaining land located on the edge of urban areas and areas with sensitive ecology
		− Controlling wind and reducing air pollution
		− Reducing noise
		− Controlling erosion and temperature fluctuations
	Achieving dynamic and growing urban spaces within cities	− Protecting open spaces and sensitive environments in and around cities
		− Utilizing surface runoff
		− Using drought-tolerant landscape plans
		− Using drainage and treatment of wastewater, wastes and chemical contaminants and creating a water recycling process

Objectives	Strategies	Policies
		– Deploying the best natural potentials in parks and green belts
		– Providing adequate public facilities and equipment under the supervision of local managements in development locations
Adopting mixed uses to enhance visibility opportunities	Creating diverse uses, with different functional scales, in population centres	– Mixing business, civic, cultural and recreational uses together in the central focus of each community
		– Establishing shopping centres for city centres and towns in such a way that they can be accessed only by a single parking lot
	Paying attention to mixing uses in urban neighbourhoods	– Establishing public spaces and facilities for daily use in the urban neighbourhood structure
		– Locating shopping centres, schools and businesses near residential neighbourhoods
	Modifying and revising the rules for the deployment of urban uses	– Providing financial incentives to create mixed uses
		– Encouraging re-development in single-use regions
		– Using improved zoning techniques (such as flexible zoning, overlay zoning and impact zoning)
Preventing urban sprawl	Less consumption of lands with natural value	– Specifying the distinct boundary between suitable lands for development and lands that need protection
		– Creating green belts of agriculture and natural paths
	Guiding development towards central and inner-city areas	– Identifying suitable sites for infill development with optimal density
		– Optimal and maximum use of infrastructure and facilities available within cities for development
		– Increasing people's awareness of the economic, environmental and social benefits of developments in central urban areas
		– Allocating legal and financial incentives
	Controlling the disordered and unrestrained growth of development	– Empowering, reinvigorating and revitalizing new cities and towns and low-density suburbs
		– Reusing brown fields

(*Continued*)

TABLE 5.26 (Continued)

Objectives	Strategies	Policies
Creation of concentrated and compact urban communities	Supporting development with optimal density	– Considering optimum density in designing infrastructure, transportation systems and land use – Constructing multi-story parking structures – Constructing housing with optimal density and mixing single-family and multi-family houses
	Correct location of uses	– Locating institutions and organizations with the urban and regional level of service in the central cores of cities – Establishing essential services at close distances to reduce car dependency and demand for parking spaces

Note

1 One of the goals of smart growth is to create housing for low-income people. When housing options are placed next to each other, households with more diverse income levels are distributed more appropriately across the city, and public services are presented more efficiently. This way, a dimension of social justice is realized.

References

ADA, D. A. (2011). Americans with disabilities act. *Title II Public Services and Transportation*.

Aickin, S. (2008). Bus Station. Flickr.

Bannister, J., Fyfe, N., & Kearns, A. (2006). Respectable or respectful? In(civility) and the city. *Journal of Urban Studies, 43*(5).

Baumgartner, D., Marjoux, D., Willinger, R., Carter, E., Neal-Sturgess, C., Guerra, L., . . . Hardy, R. (2007). *Pedestrian safety enhancement using numerical methods*. Paper presented at the Proceedings of the 20th International Technical Conference on the Enhanced Safety of Vehicles Conference, Lyon, France.

Bendick Jr, M., & Rasmussen, D. W. (1996). Enterprise zones and inner-city economic revitalization. In G. E. Peterson and C. W. Lewis (Eds.), *Reaaan and the cities* (pp. 97–129). Washington, DC: The Urban Institute.

Berechman, J., Ozmen, D., & Ozbay, K. (2006). Empirical analysis of transportation investment and economic development at state, county and municipality levels. *Transportation, 33*(6), 537–551.

Berke, P., Godschalk, D., Kaiser, E., & Rodrigerz, D. (2006). *Urban land use planning*. Chicago: University of Illinois Press.

Boyld, R. (2006). The value of civility? *Urban Studies, 43*(5/6).

Brambilla, R., & Longo, G. (1977). *For pedestrians only: Planning, design, and management of traffic-free zones*. New York: Watson-Guptill.

Breheny, M. J. (1992). *Sustainable development and urban form* (Vol. 2). London: Pion Limited.

Burden, A., Burney, D., Farley, T., & Sadik-Khan, J. (2009). *Active design guidelines promoting physical activity and health in design.* Retrieved from New York.

Campbell, T. (2013). *Beyond smart cities: How cities network, learn and innovate.* London: Routledge.

Carruthers, J. I., & Ulfarsson, G. F. (2002). Urban sprawl and the cost of public services. *Environment and Planning B.*

Chen, A., Yang, H., Lo, H. K., & Tang, W. H. (2002). Capacity reliability of a road network: An assessment methodology and numerical results. *Transportation Research Part B: Methodological, 36*(3), 225–252.

CLADCP. (2008). *Walkability checklist.* Retrieved from Los Angeles: http://urbandesignla.com/resources/docs/LAWalkabilityChecklist/lo/LAWalkabilityChecklist.pdf

Curtis, C., & James, B. (2004). An institutional model for land use and transport integration. *Urban Policy and Research, 22*(3), 277–297.

Dowling, R., Flannery, A., Landis, B., Petritsch, T., Rouphail, N., & Ryus, P. (2008). Multi-modal level of service for urban streets. *Transportation Research Record: Journal of the Transportation Research Board,* (2071), 1–7.

Elcorredor. (2009). Pixabay.com.

Evans, P. B. (2002). *Livable cities?: Urban struggles for livelihood and sustainability.* Berkeley: University of California Press.

Ewing, R. (1999). *Pedestrian and transit friendly design: A primer for smart growth.* Washington, DC: International City/County Management Association and Smart Growth Network.

Ferguson, E. (1990). Transportation demand management planning, development, and implementation. *Journal of the American Planning Association, 56*(4), 442–456.

Filion, P., & McSpurren, K. (2007). Smart growth and development reality: The difficult co-ordination of land use and transport objectives. *Urban Studies, 44*(3).

Galingan, Z. C. (2009). Pedestrian-friendly streetscape in a tropical business district. *Muhon, 1*(1), 9–15.

Gehl, J., & Soholot, H. (2002). Public spaces and public life – City of Adelaide, South Australian Government. *Planning SA/City of Adelaide/Capital City Committee/Gehl Architects.*

Gehl, Y. (2002). *Public space and public life, city of Adelaide.* Retrieved from Adelaide.

Gerilla, G., Hokao, K., & Takeyama, Y. (1995). Proposed level of service standards for walkways in Metro Manila. *Journal of the Eastern Asia Society for Transportation Studies, 1*(3), 1041–1060.

Girardet, H. (2004). The metabolism of cities. In S. M. Wheeler & T. Beatley (Eds.), *The sustainable urban development reader.* London & New York: Routledge.

Gomes Franco, D. (2007). Fused Grid: Wikipedia.org.

Grammenos, F., & Pidgeon, C. (2005). *Fused grid planning in a Canadian city.* Wharton Real Estate Review, University of Pennsylvania.

Handy, S. (2005). Smart growth and the transportation-land use connection: What does the research tell us? *International Regional Science Review, 28*(2).

Harkey, D. L., & Zegeer, C. V. (2004). *PEDSAFE: Pedestrian safety guide and countermeasure selection system.* Retrieved from Washington: https://trid.trb.org/view/753660

Heath, G. W., Brownson, R. C., Kruger, J., Miles, R., Powell, K. E., Ramsey, L. T., & Services, T. F. o. C. P. (2006). The effectiveness of urban design and land use and transport policies and practices to increase physical activity: A systematic review. *Journal of Physical Activity & Health, 3,* S55.

HosseinZadeh Delir, K, & Safari, F. (2012). The effect of smart planning on urban spatial planning. *Journal of Geography and Urban Development,* (1) (Spring and Summer).

Illinois Agency for Planning. (2010). *Go To 2040: Comprehensive regional plan.* Retrieved from Chicago, IL.

Jacobs, J. (1961). *The death and life of great American cities*. New York: Random House.

Jiang, B., & Claramunt, C. (2004). A structural approach to the model generalization of an urban street network. *GeoInformatica, 8*(2), 157–171.

Jouan, R. (2009). Los Angeles: Wikimedia.org.

Kim, J. (2007). *Testing the street connectivity of new urbanism projects and their surroundings in Metro Atlanta Region*. Paper presented at the Proceedings, 6th International Space Syntax Symposium, İstanbul.

Krueger, P. (2011). Bike path alongside with personal cars: Wikimedia.org.

La Citta Vita. (2011). Walkable: Flickr.com.

Litman, T. (2015a). *Evaluating criticism of smart growth*. Washington, DC: Victoria Transport Policy Institute.

Litman, T. (2015b). Urban sprawl costs the American economy more than $1 trillion annually: Smart growth policies may be the answer. *American Politics and Policy Blog*.

Lotfalinejad, B. (2019). Urban design sketches collection. Not Published.

Lynch, K. (1981). *A theory of good city form*. Cambridge: MIT Press.

Maclennan, L. (2019). Delightful Paris and Cornwall at last: Endellionbarge.wordpress.com.

Martin, A. (2006). *Factors influencing pedestrian safety: A literature review*. TRL Limited.

McGroarty, J. (2010). Neihoff Urban Studio – W10 January 29, 2010.

Miller, J. S., & Hoel, L. A. (2002). The "smart growth" debate: Best practices for urban transportation planning. *Socio-Economic Planning Sciences, 36*(1).

Mogliani, G. (2019). Dubai metro. Pexels.com.

Montgomery, J. (1998). Making a city: Urbanity, vitality and urban design. *Journal of Urban Design, 3*(1), 93–116.

Moreira, A. E. (2009). New York city transit authority: Wikipedia.org.

Morgan, S. (2013). Bike lane in Portland: Wikimedia.Org.

Morris, A. E. J. (2013). *History of urban form before the industrial revolution*. London: Routledge.

Morris, M. (1997). *Creating transit-supportive land-use regulations*. Chicago: Transportation Research Board.

Newman, P., & Kenworthy, J. (2006). Urban design to reduce automobile dependence in central business district. *Opolis, 2*(1).

Obeng, K., & Ugboro, I. (2003). *Organizational commitment among public transit employees: An assessment study*. Paper presented at the Journal of the Transportation Research Forum.

Papacostas, C. S., & Prevedouros, P. D. (1993). *Transportation engineering and planning*. Washington, DC: Transportation Research Board.

Rahnama, M. R., & Abbas Zadeh, G. (2006). Comparative study of distribution/compression grading in metropolitan Sydney and Mashhad. *Geography and Regional Development Magazine*.

Reese, I. (2011). *Altoona, PA: Researching smart growth principles in a shrinking city*. Pennsylvania: Pennsylvania State University Press.

Rodriguez, E., & Goerman, P. (2004). *Transportation equity act reauthorization*. Washington, DC: National Council of La Raza.

Saelens, B. E., Sallis, J. F., & Frank, L. D. (2003). Environmental correlates of walking and cycling: Findings from the transportation, urban design, and planning literatures. *Annals of Behavioral Medicine, 25*(2), 80–91.

Seifuddini, F., & Shorjah, M. (2014). *Smart planning of land use and urban transportation, a dialectical look at urban space*. Tehran: Modiran – e- Emrooz Press.

Soeharjono, D. (2010). Stores: Flickr.com.

SPUR. (2011). Protection for sidewalks: Flickr.

Stockholm City Council. (2010). *The walkable city*. Retrieved from Stockholm

Teir, R. (1993). Maintaining safety and civility in public spaces: A constitutional approach to aggressive begging. *Louisiana Law Review, 54*(2).

The Transportation Equity Act for the 21st Century, (1998). Federal transportation bill as Public Law 105-178.

Theart, A. (2007). *Smart growth: A sustainable solution for our cities?* (Master of Philosophy in Sustainable Development Planning), University of Stellenbosch, South Africa.

Trust for Public Land. (1999). *The economic benefits of parks and open space: How land conservation helps communities grow smart and protect the bottom line.* Washington, DC: Trust for Public Land.

Tunda, F. (2007). Flickr.com.

U.S. Environmental Protection Agency. (2002). *Smart growth policy database glossary.* Washington, DC: Publication EPA.

Viriyincy. (2006). Using bicycle and Bus: Flickr.

Viriyincy. (2011). Using bicycle and Bus: Flickr.

Williams, K. M., Seggerman, K. E., & Nikitopoulos, I. (2004). *Model regulations and plan amendments for multimodal transportation districts.* Retrieved from Washington: https://rosap.ntl.bts.gov/view/dot/16126

WRCG. (2012). *Guide to creating walkable communities.* Retrieved from Riverside.

Zukin, S. (1987). Gentrification: Culture and capital in the urban core. *Annual Review of Sociology,* 129–147.

CONCLUSION

Many years ago, as Frece (2008) described the scenes behind the birth of smart growth in Maryland:

> They cancelled highway bypass projects; they threatened counties that had up-zoned farmland with the loss of farmland preservation funds; and the state began intervening in local government land use decisions, often supporting Smart Growth projects that might otherwise die in the face of local opposition.

This means having smart growth as the prime approach to city development wasn't an easy decision and led to many disagreements. However, today, smart growth is not an untested theory that might be used with caution or raises local concerns about taxes; rather, in practice it has been shown that smart growth can solve many daily urban issues caused by urban sprawl. In Chapter 2 we described the negative effects of urban sprawl and provided definitions from many scholars and institutions, which saw it as an inappropriate development pattern. This chapter also described causes of urban sprawl in nine aspects, which can be seen as *urban pushes* and *suburban pulls*. The desire to ride in a personal vehicle and owning a piece of nature (such as a belvedere or a house within a garden), on one hand, and high land value in downtown or expensive neighbourhoods, on the other, can lead to horizontal growth of a city. Although talking about smart growth isn't as challenging as those initial times in Maryland, we should notice that causes of urban sprawl are never eliminated completely from our lives, and we should be aware of coercions our cities can compel on citizens by tapping into their human desires.

As we noted before, smart growth isn't a pattern for growth that stands against sprawl and leap-frogging; rather, it's a movement or massive theory that includes prior theories such as that of the compact city, infill development, transit-oriented development, etc. and one more important essence of smart growth is the multi-scalar feature of collaboration and participation in decision making. As was seen in Chapter 3, the smart growth plan can cover a city within a region. Therefore, this means everyone must participate in every aspect, from government to local citizens. Dierwechter (2017) stressed this key factor (multi-scalar) as the prime tool that made the Seattle smart growth plan as a successful experience. Chapter 3, in reviewing some experiences about planning for smart growth, revealed aspects that weren't always seen in the ten principles of Smart Growth Network. For instance, in the case of Portland, a healthy economy was highlighted as a major principle for smart growth. Taking a look at the current academic literature on financial management (Lovegreen, Riggs, Staten, Sheehan, & Pittas, 2018; Sassen, 2018; Pan & Yang, 2018) or increasing cities' income (Alibegović, Hodžić, & Bečić, 2018; Jimenez, 2018) illustrates that there is still an appetite for financial features in cities' plans. An excellent book about smart growth and business was written by Hess (2010); however, our book focused on human-based transportation, and in Chapter 3 we provided smart growth plans with an emphasis on transportation. Obviously, smart growth is not a one-size-fits-all concept, and various aspects need to be taken into consideration, based on an individual city's needs.

There are some concerns about smart growth which were described in Chapter 4. Many people still dream about living in the suburbs in a large house with a massive yard. Additionally, there is another concern about increasing land value because so few areas of land are available. However, as Ciscel (2001) showed, the new sprawled city is expensive, both in terms of investment capital and maintenance costs, and many other scholars over the past 50 years warned about the heavy costs of urban sprawl (Harvey & Clark, 1965; Brueckner & Fansler, 1983; Brueckner, Mills, & Kremer, 2001; Zhang, Miao, Zhang, & Chen, 2018; Weilenmann, Seidl, & Schulz, 2017). With failure of urban sprawl, smart growth disadvantages regarding land value can be handled, as Gabriel, Faria, and Moglen (2006) demonstrated solutions in their research, as did Grant (2009). The smart growth movement is still developing to cover more solutions for more urban issues, and as Samuel Beckett (1999) once said, Ever tried. Ever failed. No matter. Try Again. Fail again. Fail better.

This book tried to narrow down the big picture of the smart growth movement to a comprehensive checklist of generalizable and achievable goals, strategies and policies for smart growth in Chapter 5 with an emphasis on transportation within the city. Human-based transportation, as the prime goal of this book, follows the fabulous masterpieces of Jan Gehl, *Cities for People* (2013), *Life between Buildings* (2011) and *Public Spaces and Public Life* (2002), which indicate the potential for a lively city is strengthened when more people are invited to walk, bike and stay in city spaces. There is a gap in terms of smart growth and transportation. In searching for the ideal combination of those two terms, we referred to *Smart Growth Transportation Guidelines: An ITE Recommended Practice* (2003), which is similar to our book, although its principles are its outdated. We covered more strategies in terms of creating walkable communities and planning urban infrastructures and public services, along with land regulations. Few other books about smart growth and transportation include any suggested strategies or action steps; some include such as *Smart Growth and Transportation: Issues and Lessons Learned* (2002), which is a conference proceeding and includes papers addressing questions on the What? When? Where and Why? about smart growth and transportation. The book by Outwater et al. (2014) titled *Effect of Smart Growth Policies on Travel Demand* proposed a review on modelling travel demand in the context of smart growth. This book was an attempt to complement existing research on smart growth and fill the gap between theory and practice. Although there are two major limitations to this book, which we fully acknowledge.

One of the limitations is debates about hidden structures that push cities to horizontal growth. These include the powerful hands of neoliberalism and the essence of capital itself, which, for example, can be seen in the book *Rebel Cities: From the Right to the City to the Urban Revolution* by David Harvey (2012) and also *A Brief History of Neoliberalism (2007)* which remarkably illustrates the desires of capital to consume large amounts of land and turn them to shopping malls with massive parking lots. These roots of neoliberalism and capital build up an apparatus of power that has a massive ability to expand and develop everywhere (both inside and outside of cities). Awareness of how this sophisticated structure exercises power on land regulations (see *Shadows of Power: An Allegory of Prudence in Land-Use Planning* by Hillier (2003)) and how it's acting against smart growth is not investigated in this book due to its limited scope.

The other limitation refers to the philosophy aspects of smart growth theory, including epistemology and ontology. Ontology explores discourses about the nature of smart growth under the certain lens of thought. Epistemology investigates necessary and sufficient conditions for smart growth and how it can be justified. Developing those aspects of smart growth can lead to a better, more solid and sounder theory that spotlights the deficiencies of smart growth. However, this book insists on practice rather than theory, and debating those two philosophy parameters again was beyond the scope of this book.

To achieve the comprehensive checklist of generalizable and achievable goals, strategies and policies for smart growth, we considered smart growth as a combination of a set of theories, including TOD, compact city, mixed-use districts and sustainable development, but the larger "smart term" in city and regional affaires represents new challenges for researchers. In particular, the relationship between smart growth and new rising theories about public participation – for city and local scales of smart growth planning (Nabatchi & Jo, 2018) – and geo-political – for regional scale (Allmendinger, 2018) – and recent smart city initiatives, needs more research. Smart cities are much more than hooking up traffic lights with sensors, or analysing the latest apps of ambulant hipsters in search of trendy restaurants, potential dates and available parking slots, as Townsend (2013) shows many possible uses of big data in his book or Zanella et al. (2014) debates the Internet of Things for the Padova Smart City project and what it achieved. The smart city could be the next step after smart growth and raises questions: for instance, how should we be thinking through the relationship between smart growth principles and smart city possibilities going forward? Can we combine smart city with smart growth and make a smarter city? And so much work remains.

As with the notable dearth of sufficient literature about smart growth and human-based transportation, we still need more research on creating city spaces for the human scale; nevertheless, there is marvellous recent works by scholars, albeit working in different theoretical traditions (not exclusively smart growth): Matthew Carmona (Carmona, 2010; Carmona, Heath, Oc, & Tiesdell, 2012; Carmona and Punter, 2013), on designing public spaces; Jan Gehl (Gehl, 2011, 2013), about how to lead cities for humans; Hebert (1972), on highways to nowhere; Luca Bertolini (Bertolini and Le Clercq, 2003; Bertolini, 2007; Bertolini, Le Clercq, & Straatemeier, 2008), on massive work on city transportation development; Susan Handy (2005), on the connection of smart growth and transportation land use; and Tom Daniels and Yonn Dierwechter (Dierwechter, 2017; Daniels, 2001), on the investigation of smart growth on a regional scale. Without these works, this book's most important ideas would not have been possible. We can only hope that this book represents a modest contribution to this evolving body of efficacious urban scholarship.

References

Alibegović, D. J., Hodžić, S., & Bečić, E. (2018). Limited fiscal autonomy of Croatian large cities. *Lex Localis-Journal of Local Self-Government, 16.*

Allmendinger, P. (2018). Contemporary spatial governance: The making and remaking of land use planning. In M. Bevir (Ed.), *Governmentality after neoliberalism* (pp. 16–30). London: Routledge.

Beckett, S. (1999). Worstward Ho. 1983. *Nohow On:* 99–128.

Bertolini, L. (2007). Evolutionary urban transportation planning: An exploration. *Environment and Planning A, 39,* 1998–2019.

Bertolini, L., & Le Clercq, F. (2003). Urban development without more mobility by car? Lessons from Amsterdam, a multimodal urban region. *Environment and Planning A, 35,* 575–589.

Bertolini, L., Le Clercq, F., & Straatemeier, T. (2008). Urban transportation planning in transition. *Transport Policy, 15*(2), 69–72.

Brueckner, J. K., & Fansler, D. A. (1983). The economics of urban sprawl: Theory and evidence on the spatial sizes of cities. *The Review of Economics and Statistics,* 479–482.

Brueckner, J. K., Mills, E., & Kremer, M. (2001). Urban sprawl: Lessons from urban economics [with comments]. *Brookings-Wharton Papers on Urban Affairs,* 65–97.

Carmona, M. (2010). *Public places, urban spaces: The dimensions of urban design.* London: Routledge.

Carmona, M., Heath, T., Oc, T., & Tiesdell, S. (2012). *Public places-urban spaces.* London: Routledge.

Carmona, M., & Punter, J. (2013). *The design dimension of planning: Theory, content and best practice for design policies.* London: Routledge.

Ciscel, D. H. (2001). The economics of urban sprawl: Inefficiency as a core feature of metropolitan growth. *Journal of Economic Issues, 35,* 405–413.

Daniels, T. (2001). Smart growth: A new American approach to regional planning. *Planning Practice and Research, 16,* 271–279.

Dierwechter, Y. (2017). *Urban sustainability through smart growth: Intercurrence, planning, and geographies of regional development across Greater Seattle*. New York: Springer.

Force ISGT and Engineers IoT. (2003). *Smart growth transportation guidelines: An ITE recommended practice*. Washington, DC: Inst of Transportation Engrs.

Frece, J. W. (2008). *Sprawl & politics the inside story of smart growth in Maryland*. Albany: State University of New York Press.

Gabriel, S. A., Faria, J. A., & Moglen, G. E. (2006). A multiobjective optimization approach to smart growth in land development. *Socio-Economic Planning Sciences, 40*, 212–248.

Gehl, J. (2011). *Life between buildings: Using public space*. Washington, DC: Island Press.

Gehl, J. (2013). *Cities for people*. Washington, DC: Island press.

Gehl, Y. (2002). *Public space and public life*. Adelaide: Adelaide city council.

Grant, J. L. (2009). Theory and practice in planning the suburbs: Challenges to implementing new urbanism, smart growth, and sustainability principles. *Planning Theory & Practice, 10*, 11–33.

Handy, S. (2005). Smart growth and the transportation-land use connection: What does the research tell us? *International Regional Science Review, 28*.

Harvey, D. (2007). *A brief history of neoliberalism*. New York: Oxford University Press.

Harvey, D. (2012). *Rebel cities: From the right to the city to the urban revolution*. London: Verso Books.

Harvey, R., & Clark, W. A. V. (1965). The nature and economics of urban sprawl. *Land Economics, 41*.

Hebert, R. (1972). *Highways to nowhere: The politics of city transportation*. Washington, DC: Bobbs–Merrill.

Hess, E. D. (2010). *Smart growth: Building an enduring business by managing the risks of growth*. New York: Columbia University Press.

Hillier, J. (2003). *Shadows of power: An allegory of prudence in land-use planning*. London: Routledge.

Jimenez, B. S. (2018). Organizational strategy and the outcomes of fiscal retrenchment in major cities in the United States. *International Public Management Journal, 21*, 589–618.

Lovegreen, O., Riggs, D., Staten, M. A., Sheehan, P. R., & Pittas, A. G. (2018). Financial management of large, multi-center trials in a challenging funding milieu. *Trials, 19*, 267.

Nabatchi, T., & Jo, S. (2018). 6 The future of public participation. *Conflict and Collaboration: For Better or Worse, 75*.

Outwater, M., Smith, C., Walters, J., Welch, B., Cervero, R., Kockelman, K., & Kuzmyak, J. (2014). *Effect of smart growth policies on travel demand*. Transportation Research Board.

Pan, F., & Yang, B. (2018). Financial development and the geographies of startup cities: Evidence from China. *Small Business Economics*, 1–16.

Sassen, S. (2018). *Cities in a world economy*. Thousand Oaks: Sage Publications.

Solomon, N. (2002). Smart Growth and Transportation: Issues and Lessons Learned. *Conference Proceedings 32*. Maryland.

Townsend, A. M. (2013). *Smart cities: Big data, civic hackers, and the quest for a new utopia*. New York: W. W. Norton & Company.

Weilenmann, B., Seidl, I., & Schulz, T. (2017). The socio-economic determinants of urban sprawl between 1980 and 2010 in Switzerland. *Landscape and Urban Planning, 157*, 468–482.

Zanella, A., Bui, N., Castellani, A., Vangelista, L., & Zorzi, M. (2014). Internet of things for smart cities. *IEEE Internet of Things Journal, 1*, 22–32.

Zhang, C., Miao, C., Zhang, W., & Chen, X. (2018). Spatiotemporal patterns of urban sprawl and its relationship with economic development in China during 1990–2010. *Habitat International, 79*, 51–60.

INDEX

Note: Page numbers in *italics* indicate figures and in **bold** indicate tables on the corresponding pages.